AF890444

My Book

This book belongs to

Name: ______________________________

Cover Design by :
Gowri Vemuri

First Edition :
December , 2019

Author :
Gowri Vemuri

Edited by :
Raksha Pothapragada
Ritvik Pothapragada

Questions: mathknots.help@gmail.com

Dedication

This book is dedicated to:

My Mom, who is my best critic, guide and supporter.

To what I am today, and what I am going to become tomorrow,

is all because of your blessings, unconditional affection and support.

This book is dedicated to the

strongest women of my life ,

my dearest mom

and

to all those moms in this universe.

G.V.

The Virginia Board of Education and Virginia Department of Education (VDOE) have developed the Virginia Assessment Program (VAP) to measure and evaluate students' academic progress in the Standards of Learning (SOLs). The SOLs indicate Virginia's expectations for what students should know and be able to do in the subject areas of reading, writing, mathematics, science, and history/social science.

Students in grades 3-12 to take the Standards of Learning (SOL) assessments each year. Some of the tests are required by all students each year, and others are required only at specific grade levels. Additionally, with the removal of some SOL tests in recent years, the VDOE assigned the responsibility of the creation and administration of alternate, performance - based assessments on local divisions. Student scores from these tests determine a school's and the division's state accreditation and measures progress toward meeting federal targets.

Virginia Standards of Learning (SOL) tests are generally given online unless a student has an identified and documented need to be assessed using paper,pencil format. The test question format is typically multiple choice, and each test contains some technology enhanced items.

GRADE 3-8 STANDARDS OF LEARNING(SOL) TESTS

GRADE 3	GRADE 4	GRADE 5	GRADE 6	GRADE 7	GRADE 8
	VIRGINIA STUDIES				WRITING
MATH	MATH	MATH	MATH	MATH	MATH
READING	READING	READING	READING	READING	READING
		SCIENCE			SCIENCE

NOTE : VDOE is neither affiliated nor sponsors or endorses this product.

END OF COURSE STANDARDS OF LEARNING (SOL) TESTS

GRADE 9	GRADE 10	GRADE 11
ALGEBRA I	GEOMETRY	ALGEBRA II
EARTH SCIENCE	BIOLOGY	CHEMISTRY
WORLD HISTORY I	WORLD HISTORY II	VIRGINIA & US HSTORY
		WORLD GEOGRAPHY
		ENGLISH : READING
		ENGLISH : WRITING

Any Student taking one of the courses listed here is expected to take the corresponding end-of-course SOL test. The grade levels depicted here represent grade level at which students typically participate in these courses.

NOTE : VDOE is neither affiliated nor sponsors or endorses this product.

SOL Test Scoring and Performance Reports:

Standards of Learning assessments in English reading, mathematics, science and history/social science are made up of 35-50 items or questions that measure content knowledge, scientific and mathematical processes, reasoning and critical thinking skills. English writing skills are measured with a two-part assessment that includes multiple-choice items and an essay.

Student performance is graded on a scale of 0-600 with 400 representing the minimum level of acceptable proficiency and 500 representing advanced proficiency. On English reading and mathematics tests, the Board of Education has defined three levels of student achievement: basic, proficient, and advanced, with basic describing progress towards proficiency.

Performance Achievement Levels:

- The achievement levels for grades 3-8 reading and mathematics tests are: *Pass/Advanced*, *Pass/Proficient*, *Fail/Basic*, and *Fail/Below Basic*.

- The achievement levels for science tests, history tests, and End-of-Course (EOC) tests* are: *Pass/Advanced*, *Pass/Proficient*, and *Fail/Does Not Meet*.

- The EOC Writing (2010 SOL) test, EOC Reading (2010 SOL) test, and EOC Algebra II (2009 SOL) test have an achievement level of *Advanced/College Path* in place of the *Pass/Advanced* achievement level.

NOTE : VDOE is neither affiliated nor sponsors or endorses this product.

INDEX

FORMULA SHEET

1. Area of a triangle

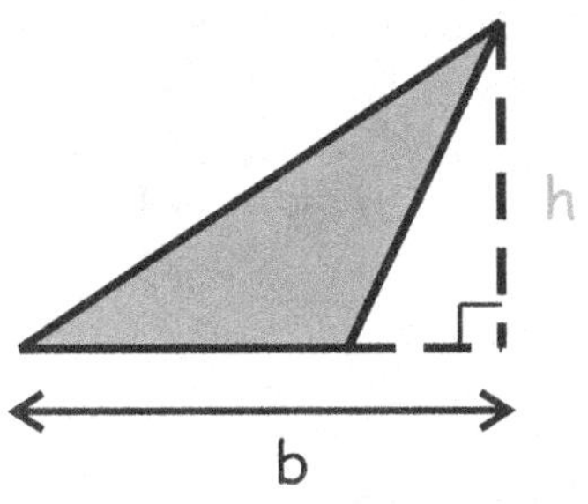

$A = \frac{1}{2} bh$

2. Area of a parellelogram

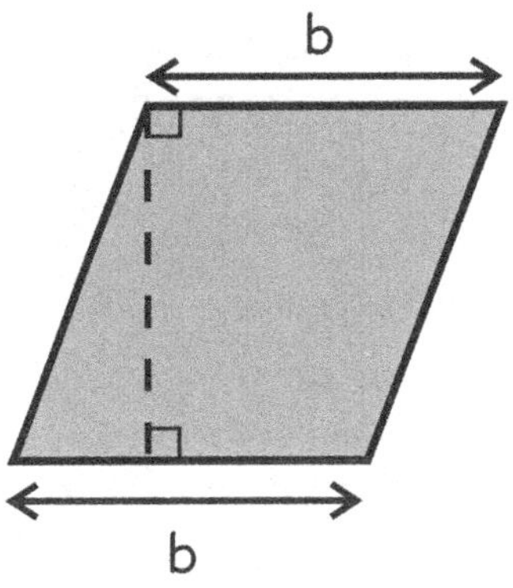

$A = bh$

3. Volume and Surface area of a cuboid

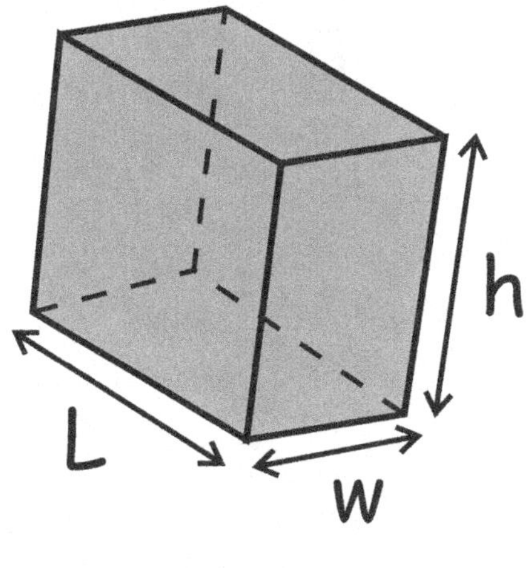

$V = lwh$

$S.A = 2(lw + lh + wh)$

4. Perimeter and Area of a Square

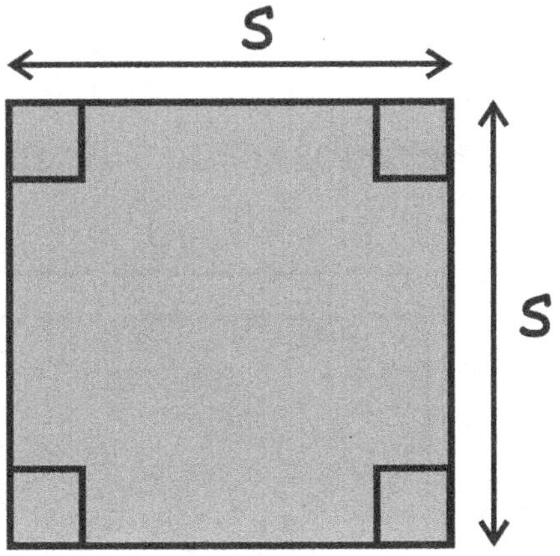

$P = 4s$

$A = s^2$

5. Area of a Trapezium

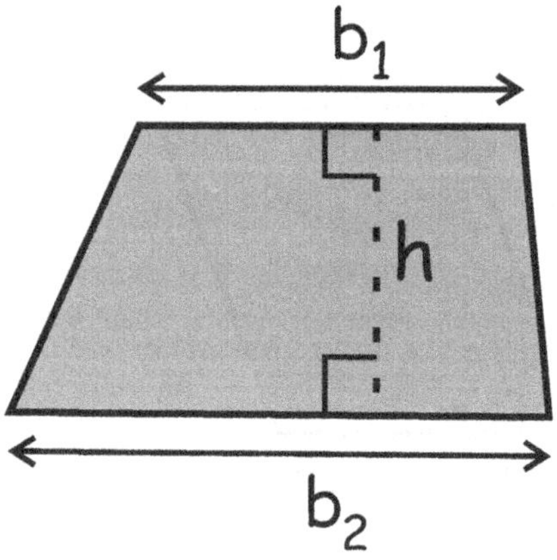

$$A = \frac{1}{2} h(b_1 + b_2)$$

6. Circumference and Area of a Circle

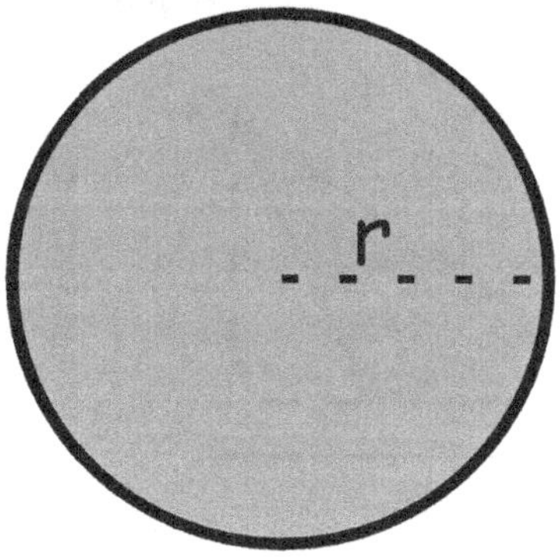

$$c = 2\pi r$$
$$A = \pi r^2$$

pi

$$\pi = 3.14$$
$$\pi = \frac{22}{7}$$

7. Perimeter and Area of a Rectangle

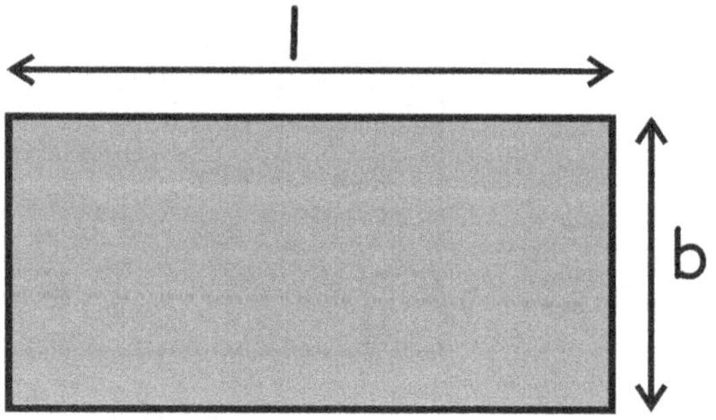

Area = l b ×

Perimeter = 2(l + b)

8. Acute angle

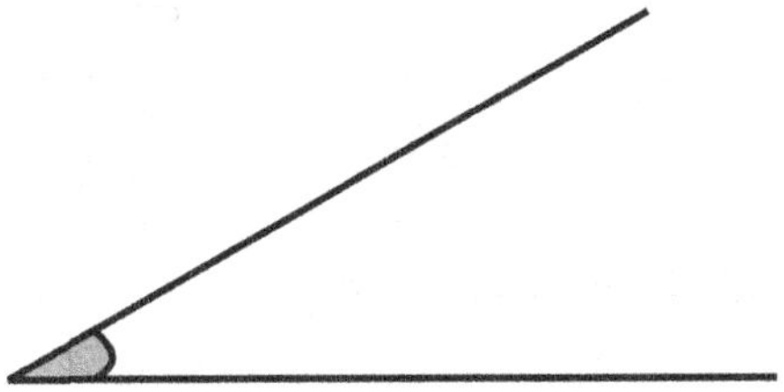

Any angle less than 90^0
is called as acute angled

9. Right angle

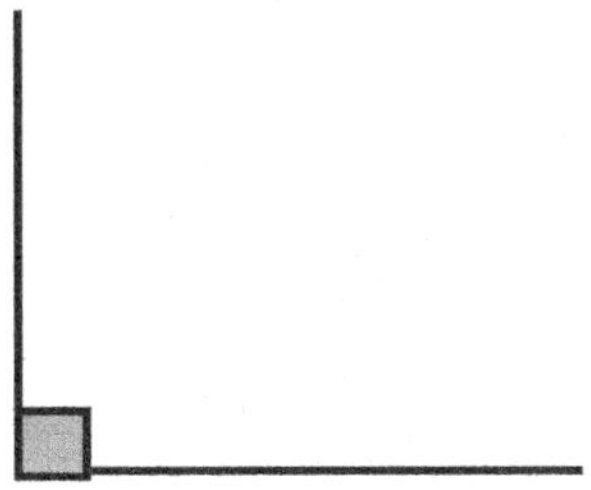

An angle equal to 90^0
is called as right angled

10. Obtuse angle

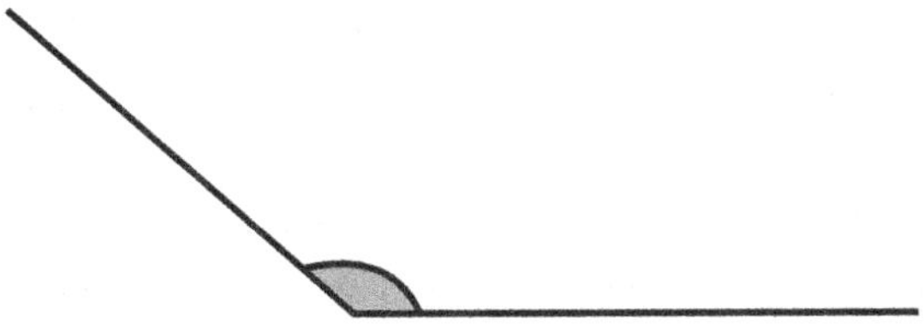

Any angle greater than 90^0
is called as obtuse angled

11. Perimeter of a polygon KLMNOP

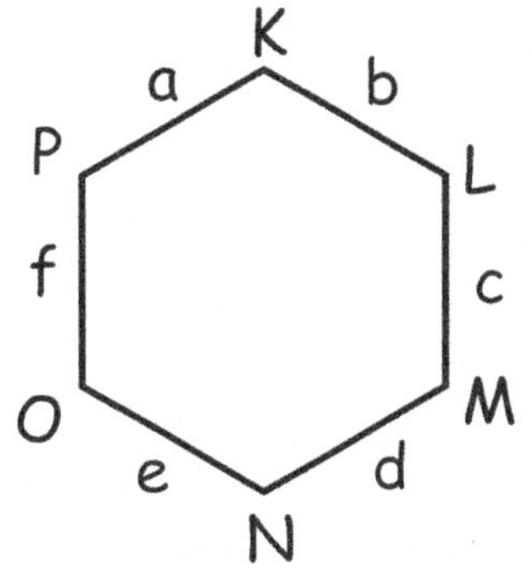

Perimeter =
(a + b + c + d + e + f) units

Abbreviations

milligram	mg
gram	g
kilogram	kg
milliliter	mL
liter	L
kiloliter	kL
millimeter	mm
centimeter	cm
meter	m
kilometer	km
square centimeter	cm^2
cubic centimeter	cm^3

year	yr
month	mon
hour	hr
minute	min
second	sec

volume	V
total Square Area	S.A
area of base	B
ounce	oz
pound	lb
quart	qt
gallon	gal.
inches	in.
foot	ft
yard	yd
mile	mi.
square inch	sq in.
square foot	sq ft
cubic inch	cu in.
cubic foot	cu ft

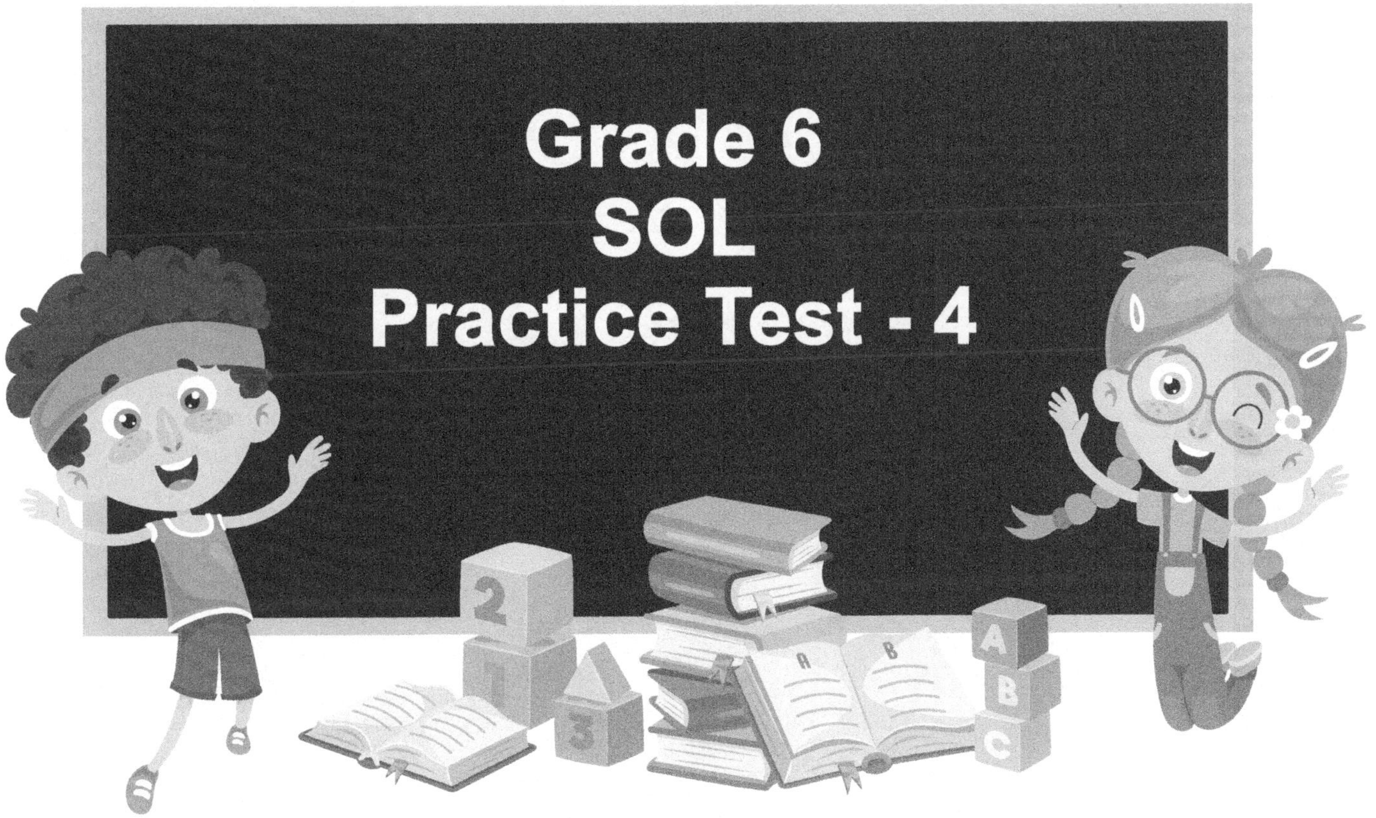
Grade 6
SOL
Practice Test - 4

1. Jack bought shoes for $99 a pair and tennis balls for $33.55 per box. To calculate the amount of money Jack spent , which information is needed ?

    (A) The number of shoes and tennis balls packs
    (B) The number of tennis balls packs
    (C) The number of pairs of socks
    (D) The number of shoes

2. Ryan bought stamps for $12.50, $19.25, and $11.20, of denominations $0.50, $0.25 and $0.10 respectively. He gave the clerk $50.00. Which is closest to the amount of change Tanya should receive ?

    (A) $7.07 (B) $6.50
    (C) $6.20 (D) $6.75

3. Evaluate 68.06 ÷ 0.04 = ?

    (A) 1701.50 (B) 1701.05
    (C) 170.65 (D) 1705.05

4. Mala made a green bead garland of length 2 ft and 11 inches and a gold bead garland of 1 ft and 5 inches and joined them together. How many inches of garland did she made ?

    (A) 52 inches (B) 4 ft 52 inches
    (C) 55 inches (D) 48 inches

5. Eric sells his small art pieces for \$11.70 each, Big Art pieces for \$46.50 each. He sold 5 small and two big art pieces today. Which is closest to the total amount he got through sales today ?

(A) \$152
(B) \$55
(C) \$150
(D) \$60

6. Evaluate $11.4 \div 28.5 = ?$

(A) 24.5
(B) 12.2
(C) 2.5
(D) 0.4

7. Evaluate $\frac{19}{13} \div \frac{7}{5} = ?$

(A) $1\frac{4}{91}$
(B) $\frac{4}{65}$
(C) $1\frac{3}{65}$
(D) $2\frac{56}{65}$

8. What is the greatest common factor of

51 , 68

(A) 6
(B) 17
(C) 68
(D) 204

9. Johnson works at a sandwich shop and he made the below table for todays sales.

ITEM	Number of sandwiches sold	Total sale amount
Cheese sandwich	70	$ 350
Chicken sandwich	105	$ 787.5
Ham Sandwich	90	$ 810

Based on the data in the table above , Which could be used to determine the cheese sandwich cost before tax ?

(A) Multiply the total sale amount by the total number of sandwiches sold

(B) Divide the total sale amount by the total number of sandwiches sold

(C) Multiply the total sale amount by the number of cheese sandwiches sold

(D) Divide the total sale amount of cheese sandwiches by the number of cheese sandwiches sold

10. Find the percentage equivalent of $\frac{97}{125}$

(A) $97\frac{1}{8}$ %

(B) $77\frac{3}{5}$ %

(C) $117\frac{3}{5}$ %

(D) $\frac{97}{125}$ %

11. Which of the below is the composite number ?

(A) 997

(B) 573

(C) 601

(D) 241

12. Angle measure closest to 94^{0}

(A)

(B)

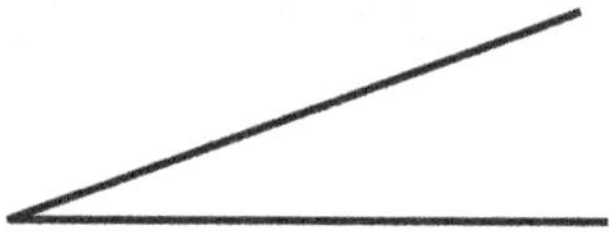

(C)

(D)

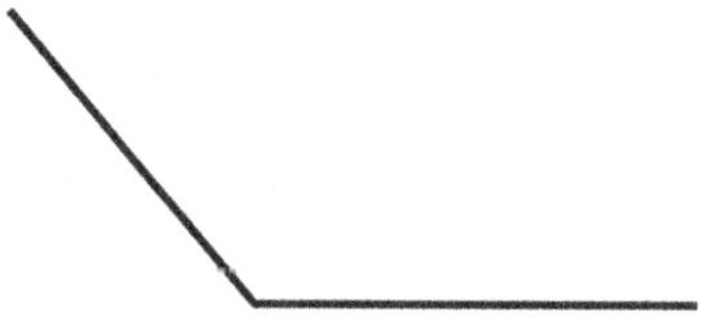

13. Which of the below statements is true ?

(A) $20 > 35$

(B) $-44 > 44$

(C) $-900 < 900$

(D) $-15 > 11$

14. Even number is ?

(A) Units digit is always 2.
(B) Divisible by 2
(C) Always divisible by 4
(D) Units digit is always 0

15. Which of the below decimals are ordered from greatest to least ?

(A) 6.976 , 6.665, 6.56 , 6.147 , 6.07
(B) 665 , 6.56 , 6.147 , 6.976 , 6.07
(C) 6.976 , 6.07, 6.665 , 6.56 , 6.147
(D) 6.147 , 6.976 , 6.665 , 6.56 , 6.07

16. Kate's apple pie recipe needs $4\frac{3}{4}$ lbs of apples. Her friend gave $3\frac{3}{4}$ lbs of apples from her farm. How many more apples are needed to make the apple pie ?

(A) $3\frac{5}{8}$
(B) $\frac{5}{7}$
(C) $2\frac{3}{7}$
(D) 1

17. How many yards are equivalent to 21 feet ?

(A) 6.3 yards
(B) 7 yards
(C) 9.2 yards
(D) 6.5 yards

18. Which is closest to the area of a Black Board ?

(A) 1600 square feet (B) 8 square inches

(C) 48 square feet (D) 480 feet

19. A turning wheel has a diameter of 4 meters. Which is closest to the circumference of this wheel ?

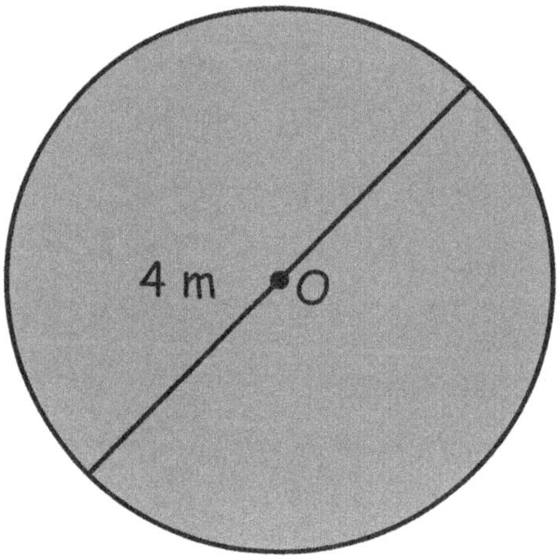

(A) 23∏ m (B) 46∏ m

(C) 22∏ m (D) 4∏ m

20. How many kilograms are equivalent to 11 pounds ? (Approximately)

(A) 3 kilo grams (B) 5 kilo grams

(C) 6 kilo grams (D) 12 kilo grams

21. A toy box contains 11 blue cars,2 red cars and 6 yellow cars. One car is randomly picked from the box. What is the probability that is a yellow car ?

(A) $\frac{5}{20}$ (B) $\frac{9}{20}$

(C) $\frac{1}{20}$ (D) $\frac{6}{19}$

22. Find the arithmetic mean of first five multiples of 4.

(A) 16 (B) 4

(C) 12 (D) 20

23. Jim sold a number of apartments from January to March as shown below

Stem	Leaf
26	0 0 5
27	5 5 8
28	1 1 2 2 9
29	3 3 3 3 4 8
30	5 5

Key :
1|5 = 15

Which of the following is a *true* conclusion based on the data in the stem-and-leaf plot ?

(A) Four houses are sold at a price of $293k.
(B) Highest number of houses are sold for $300k.
(C) All sold house prices are below $150k.
(D) The least price of the house sold was $195k.

24. Andy scored the below points in his tennis tournament during his spring break. Based on this data he drew a box and whisker plot. Chose the correct option from the below.

4	4	5	5	6	6	6	7	7	7	9

(A)

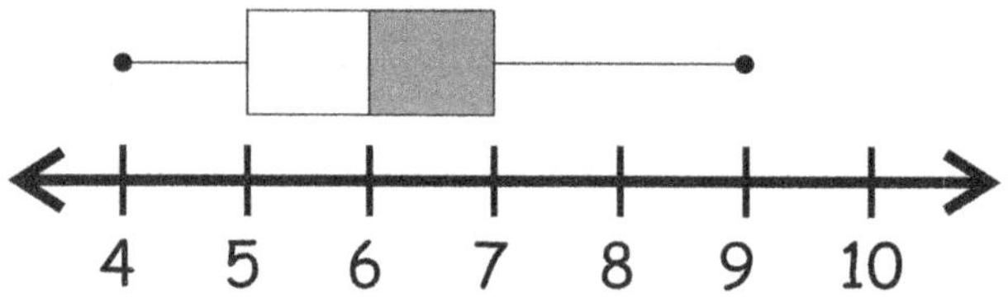

(B)

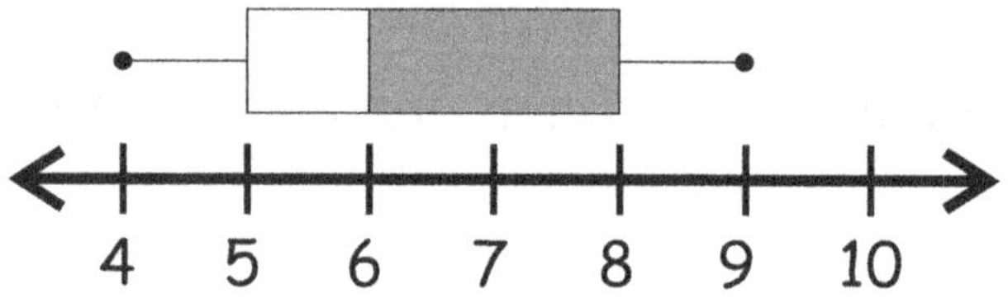

(C)

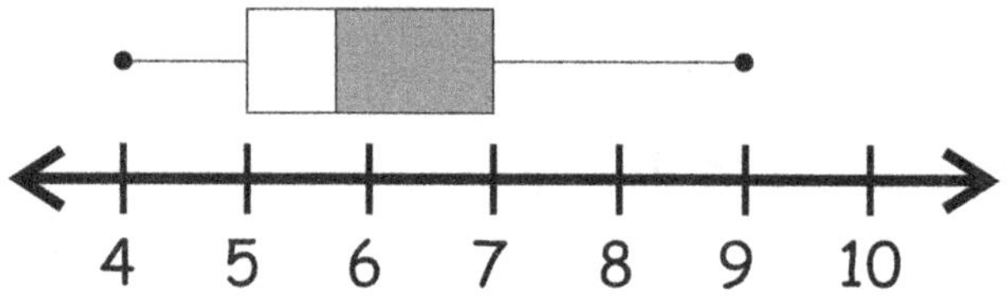

(D)

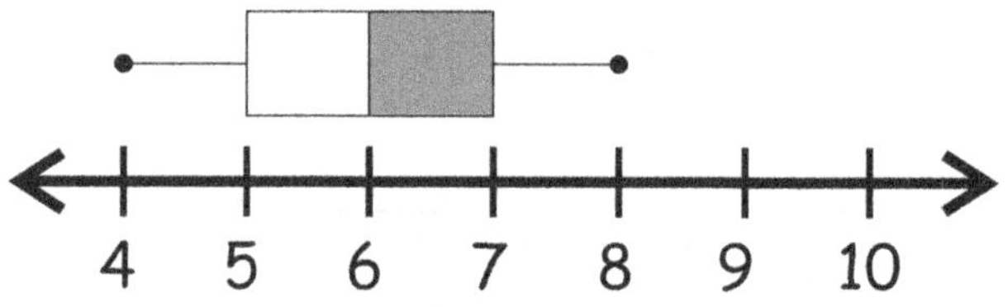

25. Angela sells 2 types of desserts with two choices of toppings at her bakery.

Deserts	Topping 1	Topping 2
Cup cakes	Vanilla	Sprinkles
Brownie	chocolate	M and M's

Based on the above chart, which of the following shall show all possible dessert combinations at Angela's store , if a customer wants to pick 1 dessert with topping 1 and topping 2 ?

(A) 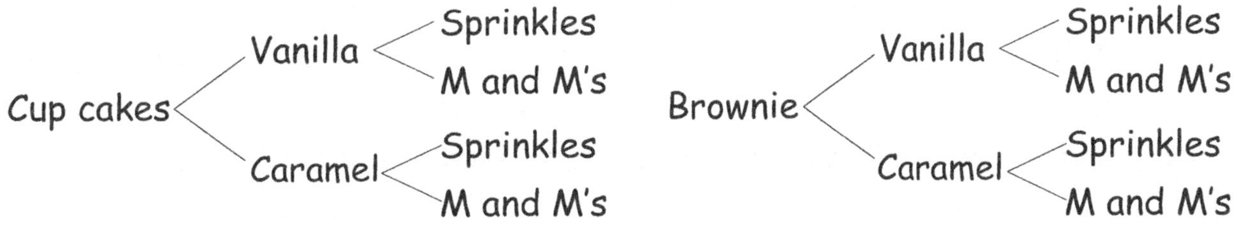

(B) 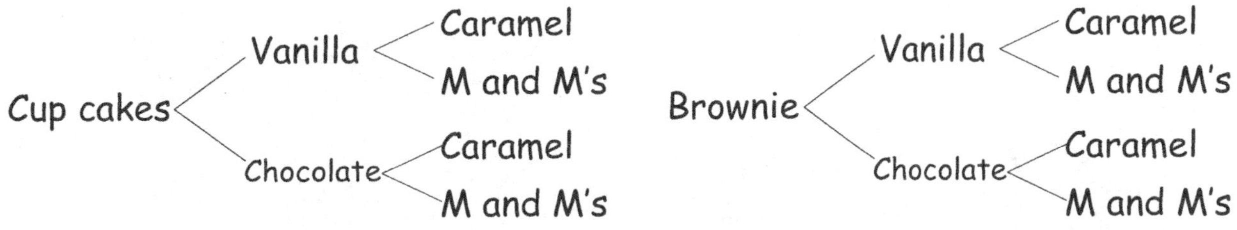

(C) 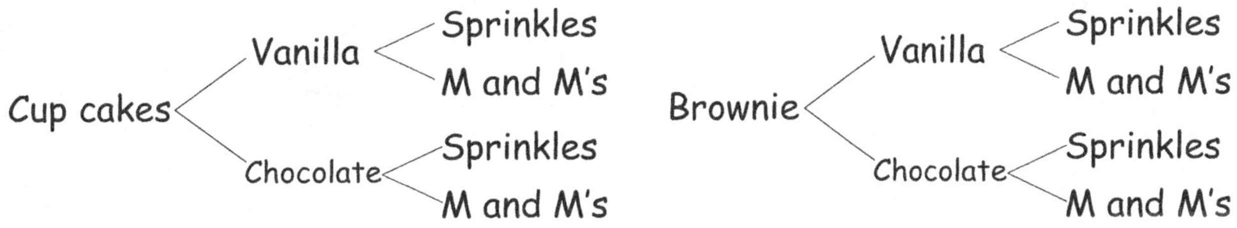

(D) 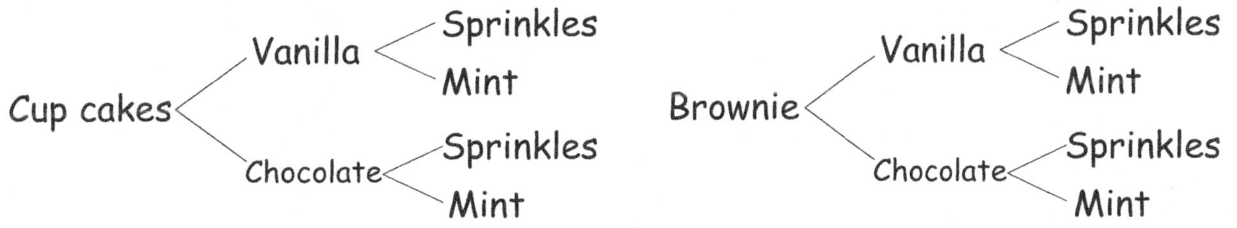

26. The below number sentences true for what value of "n" ?

$$n + 24 = 65$$

(A) $2\frac{17}{24}$ (B) 7

(C) 89 (D) 41

27. which word best describes the circled number

$$\boxed{-11}\,m = 209$$

(A) Variable (B) Co-Efficient

(C) Term (D) Equation

28. Based on the below equation. Find is the value of of Z

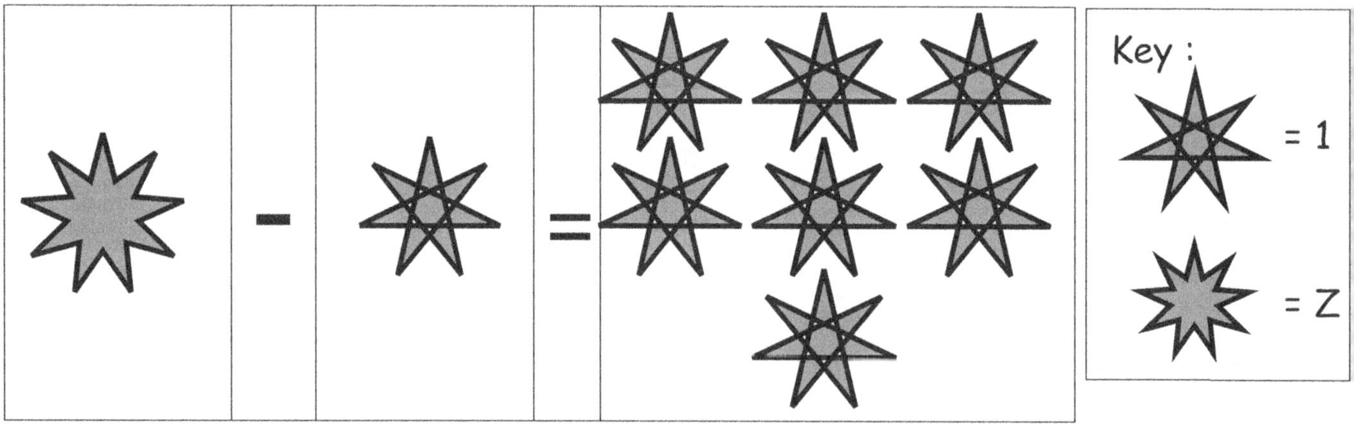

(A) 6 (B) 5

(C) 1 (D) 8

29. Find the solution of the below equation ?

$$14 = \frac{a}{6}$$

(A) $2\frac{1}{3}$ (B) 84

(C) 8 (D) 20

30. Find the 7th number in the Arithmetic pattern

31 , 43 , 55 , 67 , ...

(A) 105 (B) 91

(C) 103 (D) 87

31. Which of the below is not a perfect square ?

(A) 16 (B) 9

(C) 8 (D) 4

32. Evaluate the below :

$$3^2.3^2$$

(A) 3^3 (B) 3^2

(C) 3^5 (D) 3^4

33. What is the 18th figure in the below pattern ?

12312312...

(A) 2 (B) 3

(C) 1 (D) 12

34. Find the number sentence modeled by the shapes shown ?

Key :

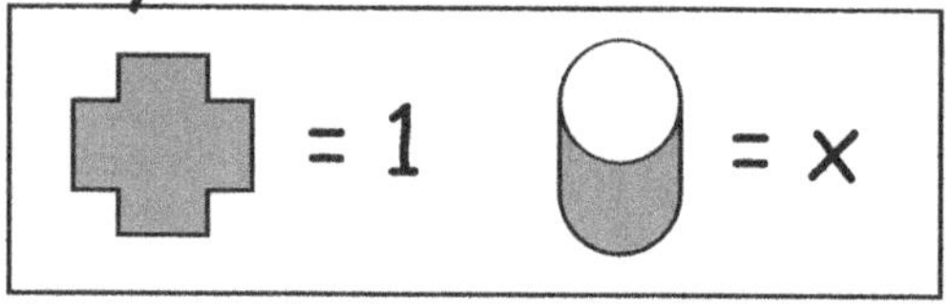

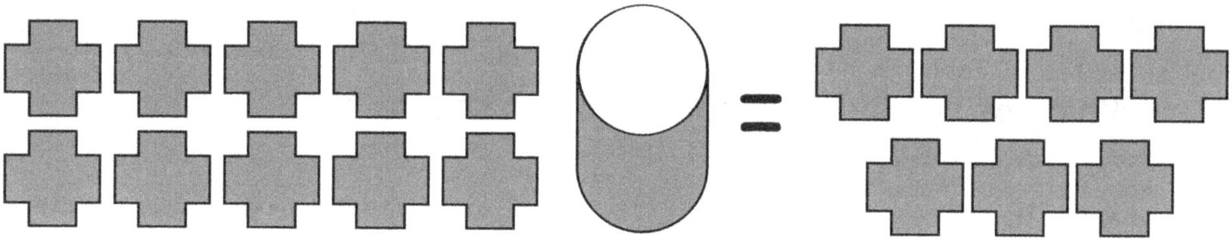

(A) $x + 7 = 10$ (B) $7x = 10$

(C) $10x = 7$ (D) $x - 7 = 10$

35. What is the 7th term in the pattern given below ?

21, 34 , 55 ,89 , 144,.....

(A) 377 (B) 233

(C) 103 (D) 87

36. The below pattern is based on a rule. Which of the below options defines the rule?

4096 , 1024 , 256 , 64 , ...

(A) Add 44 to the previous number.
(B) Multiply the previous number by 4
(C) Subtract the previous number by 40
(D) Divide the previous number by 4

37. The variable in the below expression is ?

$$\frac{22x}{5} = 99$$

(A) x
(B) 100
(C) 22
(D) $\frac{22}{5}$

38. Find the area of the rectangle ABCD given below

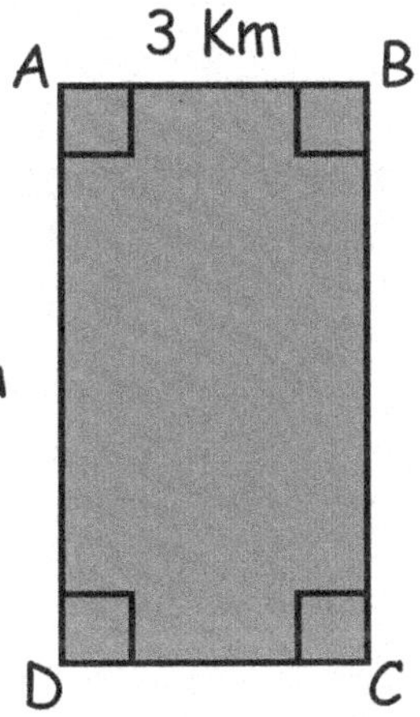

(A) 48 km^2
(B) 12 km^2
(C) 16.1 km^2
(D) 24 km^2

39. How many meters equal to 5 kilometers ?

(A) 50000 (B) 5000

(C) 500 (D) 1000

40. Find the missing angle in the given quadrilateral ABCD as given below

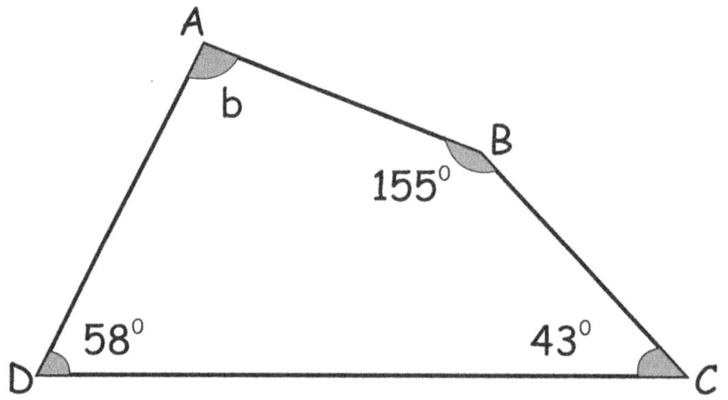

(A) 119° (B) 111°

(C) 104° (D) 118°

41. To find which of the below we need the most repeating value of data set ?

(A) Mode (B) Range

(C) Mean (D) Median

42. Find the area of the circle given below ? (Approximately)

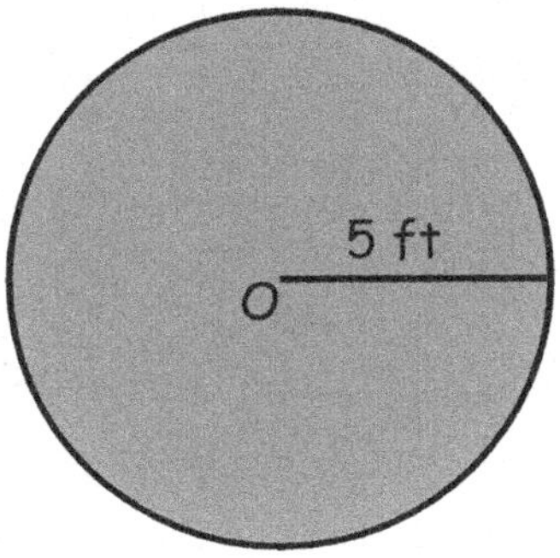

(A) 79 ft^2 (B) 13 ft^2

(C) 99 ft^2 (D) 18 ft^2

43. Which of the below represents all the factors of 78 ?

(A) 1 , 2 , 3 , 6 , 26 , 39 , 78
(B) 1 , 2 , 6 , 10 , 13 , 39 , 78
(C) 1 , 2 , 3 , 6 , 13 , 26 , 39 , 78
(D) 1 , 2 , 3 , 6 , 13 , 26 , 78

44. What is the least common multiple of

18 , 24

(A) 18 (B) 6

(C) 432 (D) 72

45. Find the median of the data given below

4 , 4 , 7 , 11 , 17 , 20 , 35 , 65 , 81 , 104

(A) 18.5 (B) 10

(C) 11 (D) 7

46. Evaluate $3\frac{1}{2} \times \frac{2}{9} \times \frac{2}{3}$ = ?

(A) $\frac{5}{27}$ (B) $\frac{14}{27}$

(C) $\frac{37}{54}$ (D) $8\frac{1}{2}$

47. Find the range of the below data set

40 , 57 , 61 , 42 , 55 , 63 , 71 , 71 , 64 , 36

(A) 47 (B) 40

(C) 31 (D) 35

48. Identify the figure below

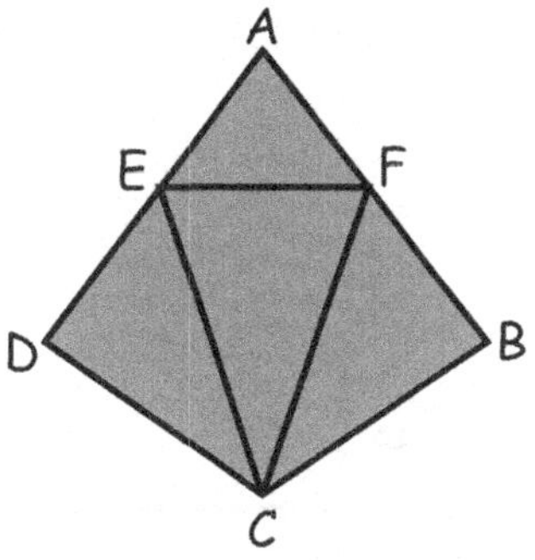

(A) Cone
(B) Pentagonal Prism
(C) Triangular pyramid
(D) Square prism

49. Sales of "Brand A" refrigerator across various years is shown as below

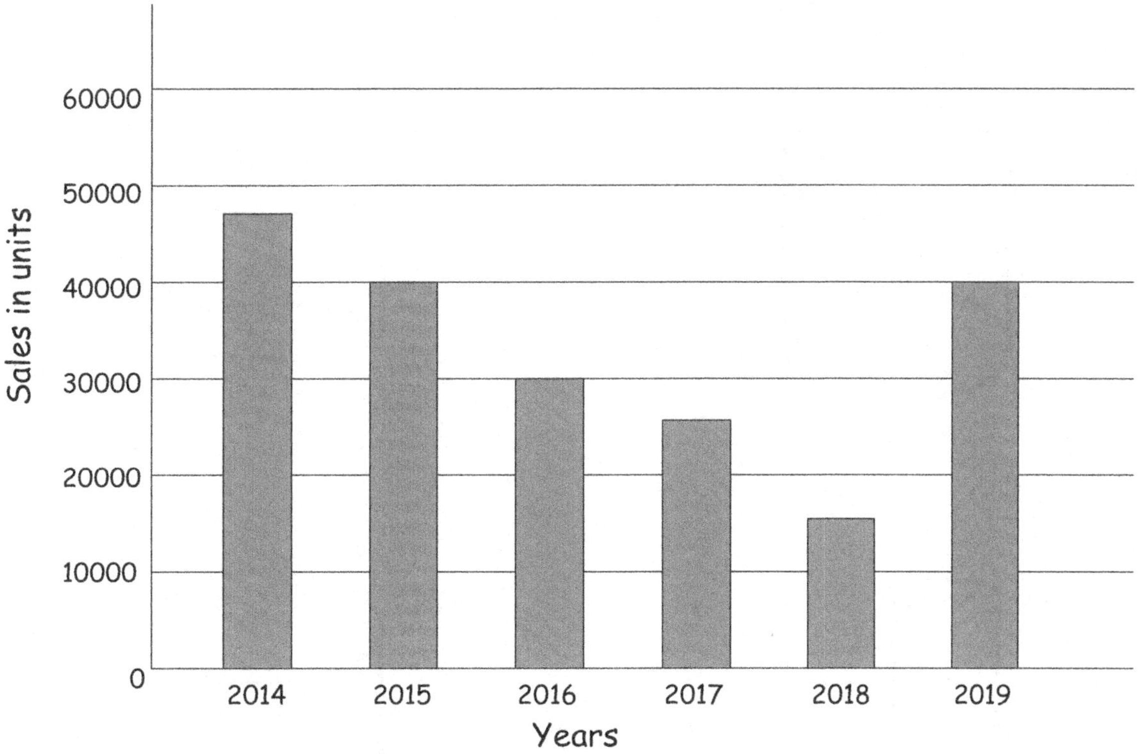

The difference in the sales of cellular phones for the years 2014 and 2016 is ?

(A) 500 units
(B) 1000 units
(C) 5000 units
(D) 18000 units

50. Find the net of the 3D figure given below

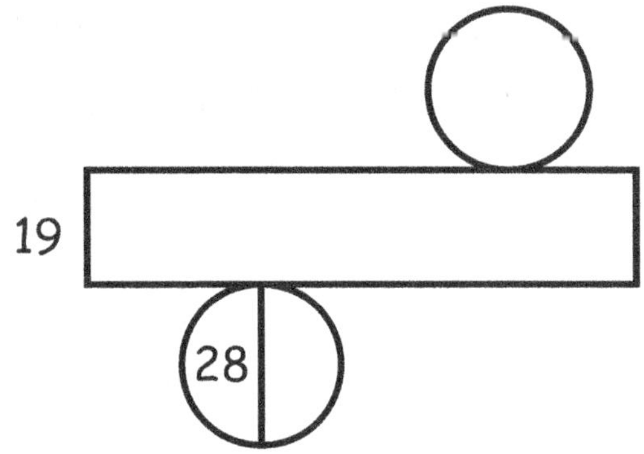

(A)

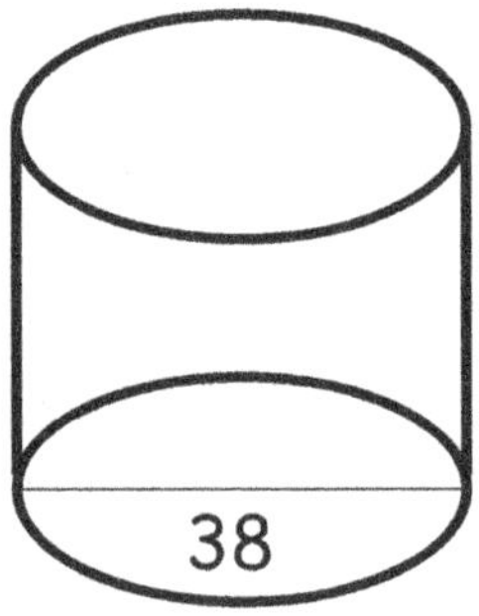

(B)

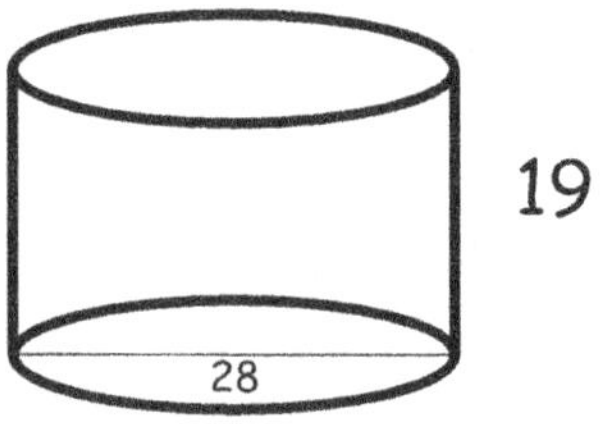

(C)

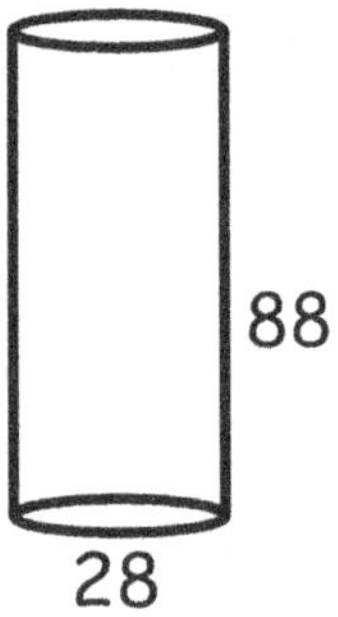

(D)

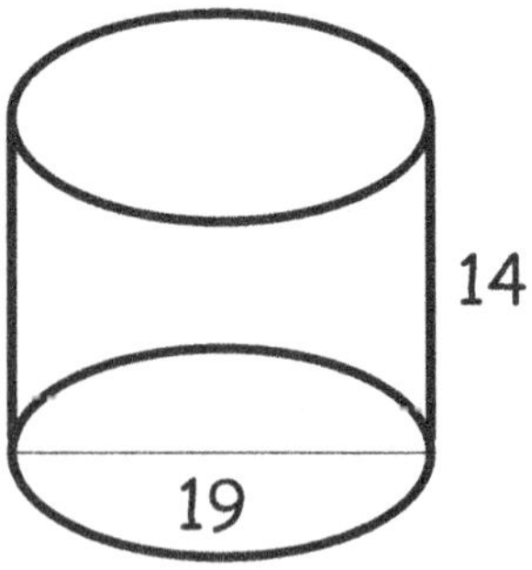

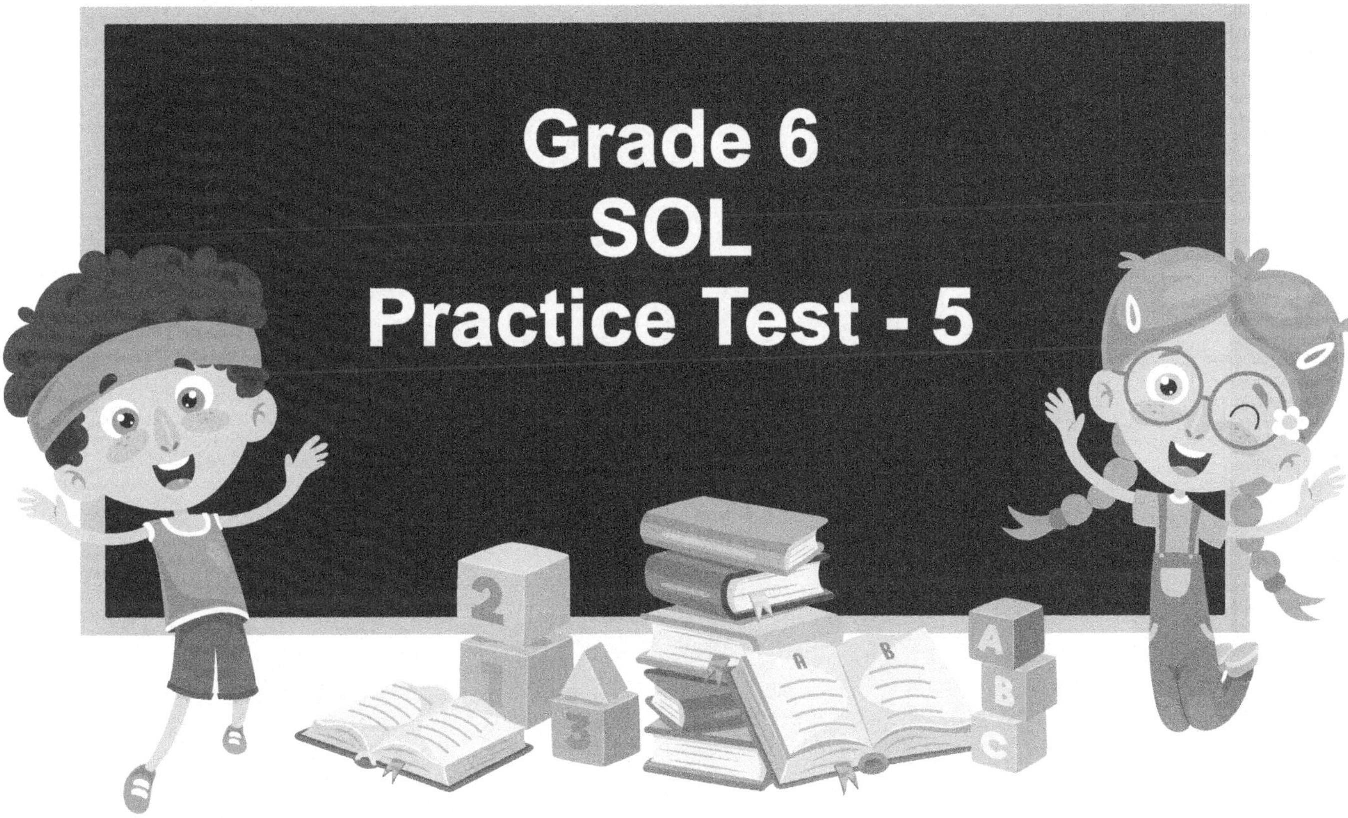
Grade 6
SOL
Practice Test - 5

1. Tim sold brand-X cars for $45,500 each and brand-Y cars for $105,000 each this week. What information is needed to find this week's sales amount by Tim ?

(A) The number of brand-X cars sold
(B) The number of brand-Y cars sold
(C) The number of brand-X cars and brand-Z cars sold
(D) The number of brand-X cars and brand-Y cars sold

2. Rik spent $20 for each movie ticket and buys drinks and popcorn which is $12.05 for each of his two friends and himself .The price includes sales tax. His parents gave $100 to spend. Which is closest to the amount Rik has after spending at the movie theater ?

(A) $3.85 (B) $3.75

(C) $2.95 (D) $2.97

3. Evaluate 8.450 ÷ 1.3 = ?

(A) 65.05 (B) 650

(C) 6.50 (D) 0.605

4. Seema is baking a chocolet cake and needs $3\frac{3}{4}$ pounds of flour and $1\frac{1}{4}$ pounds of cocoa powder. How many pounds of ingredients all together are required to bake the cake ?

(A) 5 pounds (B) $1\frac{3}{4}$ pounds

(C) $\frac{1}{4}$ pounds (D) $\frac{1}{5}$ ounces

5. Erin buys ten movie tickets at \$7.25 each and eight soda and popcorn combo at \$ 8.50 each. Which is closest to the total amount Erin paid at the theater ?

(A) \$138 (B) \$16

(C) \$141 (D) \$16.50

6. Evaluate $29.3 \div 0.4 = ?$

(A) 73.25 (B) 11.72
(C) 28.9 (D) 22.6

7. Evaluate $\frac{11}{22} \div \frac{4}{7} = ?$

(A) $1\frac{29}{48}$ (B) $\frac{5}{47}$

(C) $1\frac{3}{48}$ (D) $2\frac{3}{47}$

8. What is the greatest common factor of

52 , 32

(A) 416 (B) 8

(C) 4 (D) 16

9. Kevin works at a pizza shop. The total sales over the weekend are given as below.

ITEM	Number of pizzas sold	Cost per pizza
Small pizza	40	$ 10
Medium pizza	55	$ 13.99
Large pizza	100	$ 15.99

Based on the data in the table above, Which could be used to determine the total amount of sales before tax ?

(A) Multiply the cost per item by the number of items, then multiply the product

(B) Multiply the cost per item by the number of items, then add the products

(C) Divide the cost per item by the number of items, then add

(D) Divide the cost per item by the number of items, then multiply the quotients

10. What is the percentage equivalent of $\frac{123}{125}$?

(A) $\frac{123}{125}$ %

(B) $104\frac{2}{5}$ %

(C) $123\frac{1}{2}$ %

(D) $98\frac{2}{5}$ %

11. Which of the below is the composite number ?

(A) 883

(B) 114

(C) 599

(D) 677

12. Angle measure closest to 25^0

(A)

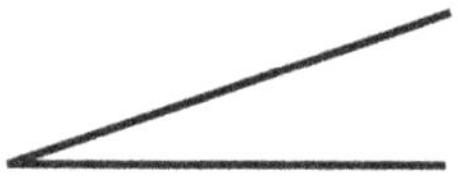

(B)

(C)

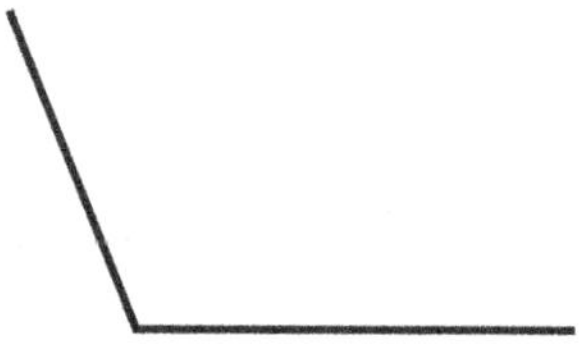

(D)

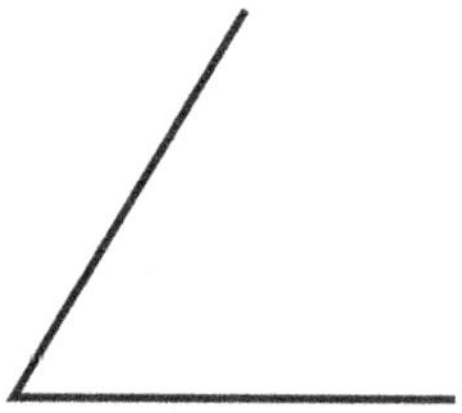

13. Which of the below statements is true ?

(A) -491 < 181

(B) -93 > 57

(C) 54 > 23

(D) -328 > 327

14. Odd number is ?

(A) Units digit is always 1.
(B) Divisible by 5
(C) Always divisible by 3
(D) Units digit is not even or zero

15. Which of the below decimals are ordered from least to greatest ?

(A) 0.224 , 0.229 , 0.374 , 0.6 , 0.97
(B) 0.224 , 0.229 , 0.97 , 0.374 , 0.6
(C) 0.97 , 0.374 , 0.224 , 0.229 , 0.6
(D) 0.6 , 0.97 , 0.229 , 0.374 , 0.224

16. Tony ran $4\frac{1}{6}$ miles today morning and $3\frac{1}{4}$ miles in the evening. How many miles did he ran today ?

(A) $9\frac{1}{6}$ (B) $9\frac{3}{5}$

(C) $7\frac{5}{12}$ (D) $7\frac{7}{12}$

17. How many inches are equivalent to 5 yards ?

(A) 190 inches (B) 180 inches

(C) 125 inches (D) 179 inches

18. Which is closest to the area of a Twin Bed ?

(A) 38 x 75 square inches
(B) 28 square inches
(C) 38 x 75 square feet
(D) 560 feet

19. A circle has a diameter of 10 miles. Which is closest to the circumference of this wheel ?

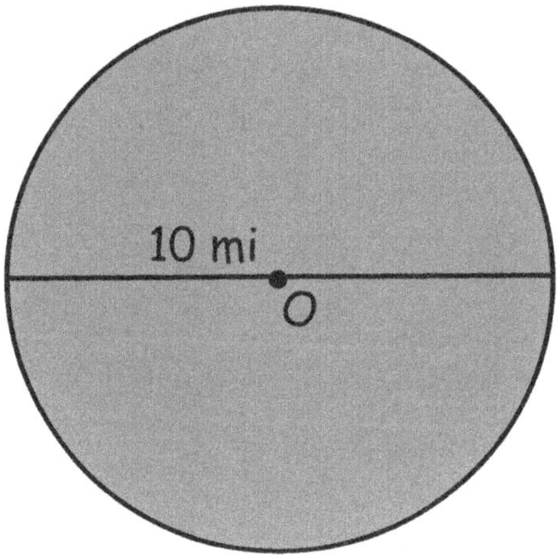

(A) 5π mi
(B) 4π mi
(C) 12π mi
(D) 10π mi

20. How many feet are equivalent to 11 meters ? (Approximately)

(A) 33 feet
(B) 44 feet
(C) 36 feet
(D) 52 feet

21. A box contains 5 vowels , 21 consonants. One Alphabet is randomly picked from the box. What is the probability that is a consonant ?

(A) $\frac{5}{21}$ (B) $\frac{21}{26}$

(C) $\frac{1}{21}$ (D) $\frac{1}{5}$

22. The scores obtained by a few students in a mathematics test are given below. Find their mean score.

56 , 50 , 76 , 45 , 40 , 62 , 34 , 56 , 45 , 64 , 53 , 42 , 56 , 45 , 56

(A) 51 (B) 50

(C) 52 (D) 53

23. The monthly expenses of steve across various weeks are given below

Stem	Leaf
2	6 8 9 9
3	4 4 4 6
4	0 0 2 3 4 8
5	4 5

Key :
1|5 = 15

Which of the following is a *true* conclusion based on the data in the stem-and-leaf plot ?

(A) Minimum expense of a week was $40.
(B) The expense amount was same across the weeks.
(C) Steve consistantly spent $25
(D) The maximum expense was $55

24. Grade 6 students sold a number of candles and raised funds for school PTA . The sales data is shown below. Based on this data the grade 6 class drew a box and whisker plot.
Chose the correct option from the below.

44	42	45	45	49	53	53	54	55	57

(A)

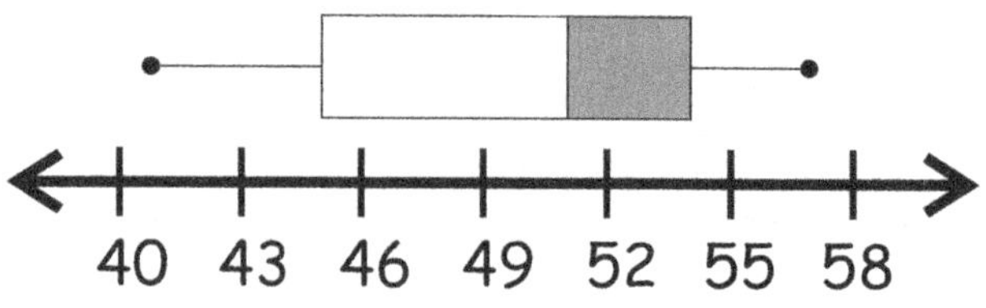

(B)

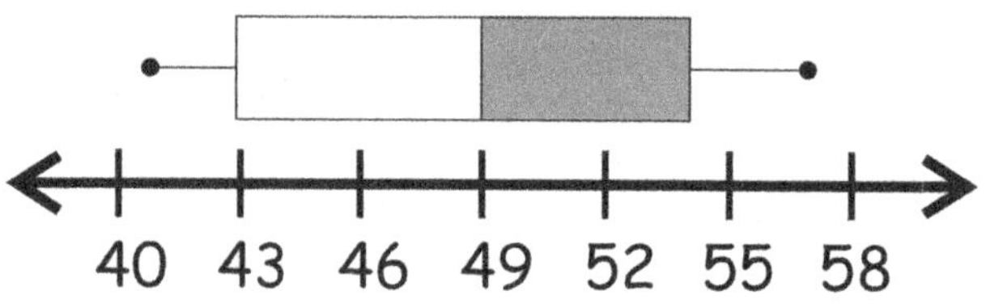

(C)

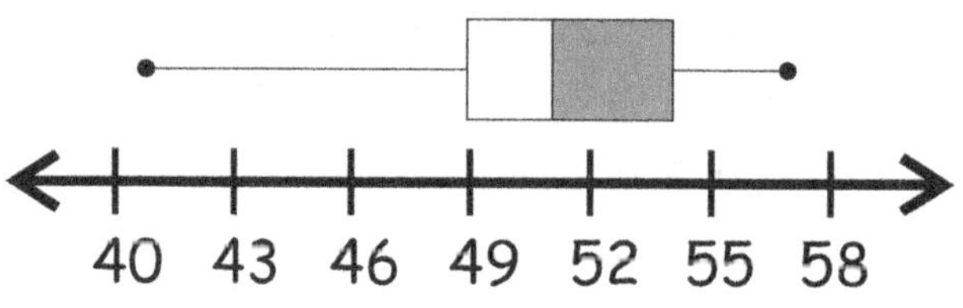

(D)

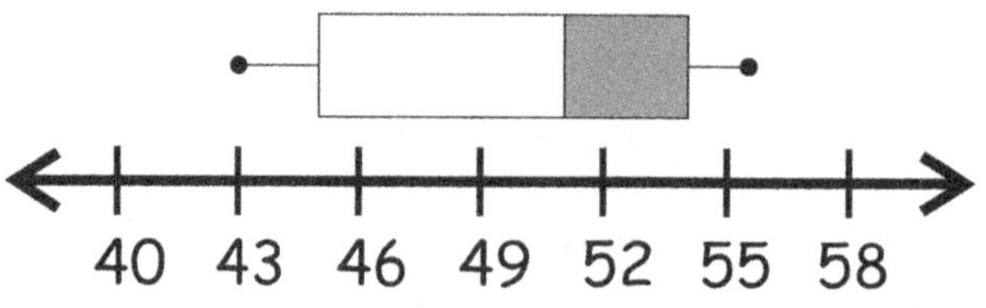

25. The famous Washington DC burger shop offers the below choices to the customers

Burgers	Type of cheese	Patty
Small	Swiss	Veggie
Medium	Chedder	Chicken

Based on the chart, which of the following shows all of Andy's possible clothing combinations of 1 color of shirt, 1 type of pants and 1 type of pants ?

(A)

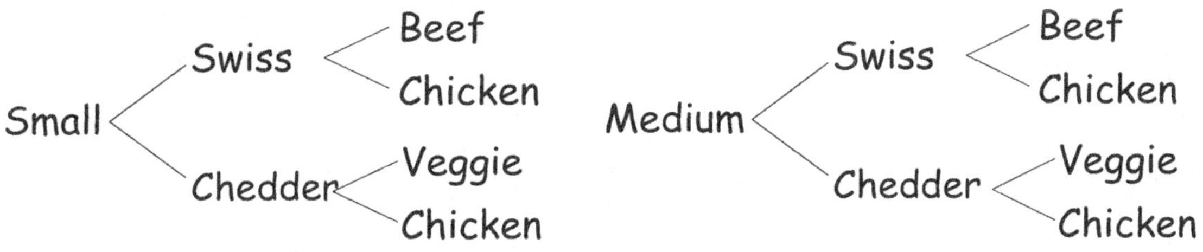

(B)

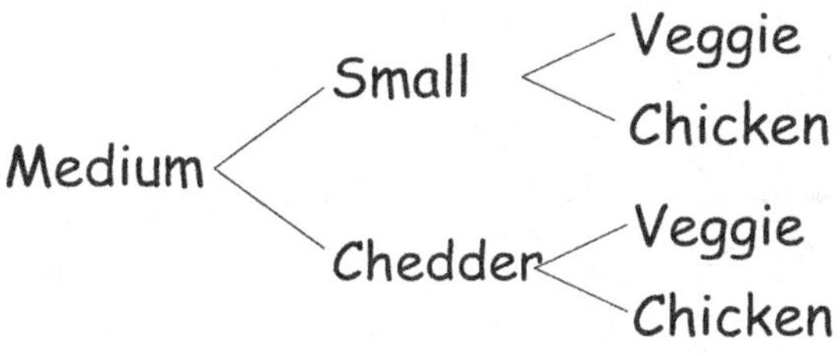

(C)

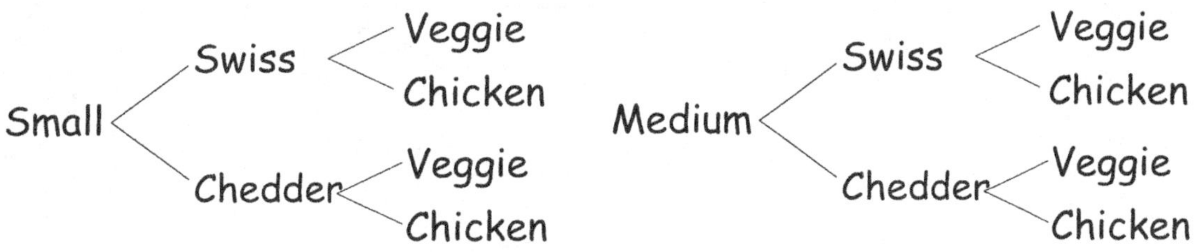

(D)

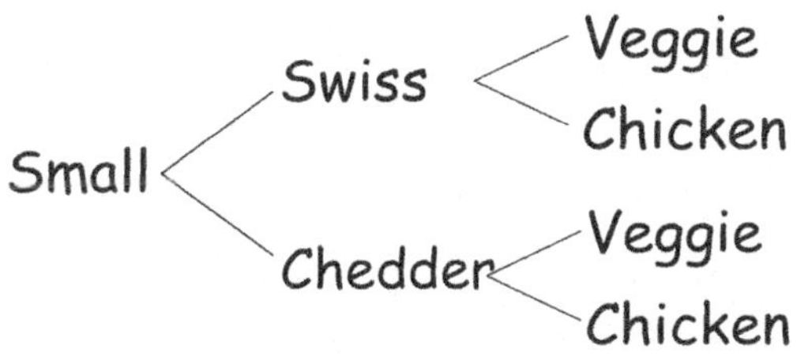

26. The below number sentences true for what value of "b"

$$b + 39 = -7$$

(A) -46 (B) 32

(C) $\frac{-7}{39}$ (D) 12

27. Which word best describes the circled number

$$\textcircled{105} - 7n$$

(A) Variable (B) Co-Efficient

(C) Term (D) Equation

28. Based on the below equation. Find is the value of Z

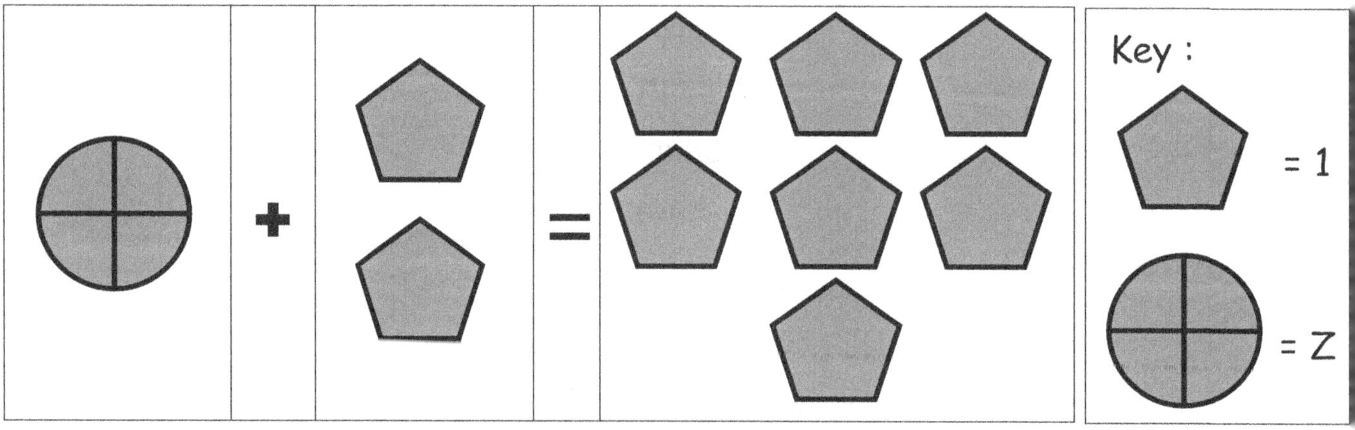

(A) 4 (B) 8

(C) 2 (D) 5

29. Find the solution of the below equation ?

$$\frac{n}{15} = -9$$

(A) -135
(B) -24
(C) 6
(D) $\frac{-3}{5}$

30. Find the 8th number in the Arithmetic pattern

212 , 227, 242 , 257 , ...

(A) 317
(B) 302
(C) 272
(D) 297

31. Which of the below is not a perfect square ?

(A) 225
(B) 289
(C) 291
(D) 196

32. Evaluate the below :

$$7^8.7^3.7^4$$

(A) 7^8
(B) 7^3
(C) 7^{15}
(D) 7^{11}

33. What is the 16th figure in the below pattern ?

1234 1234 123...

(A) 4 (B) 2

(C) 1 (D) 23

34. what number sentence is modeled by the shapes shown ?

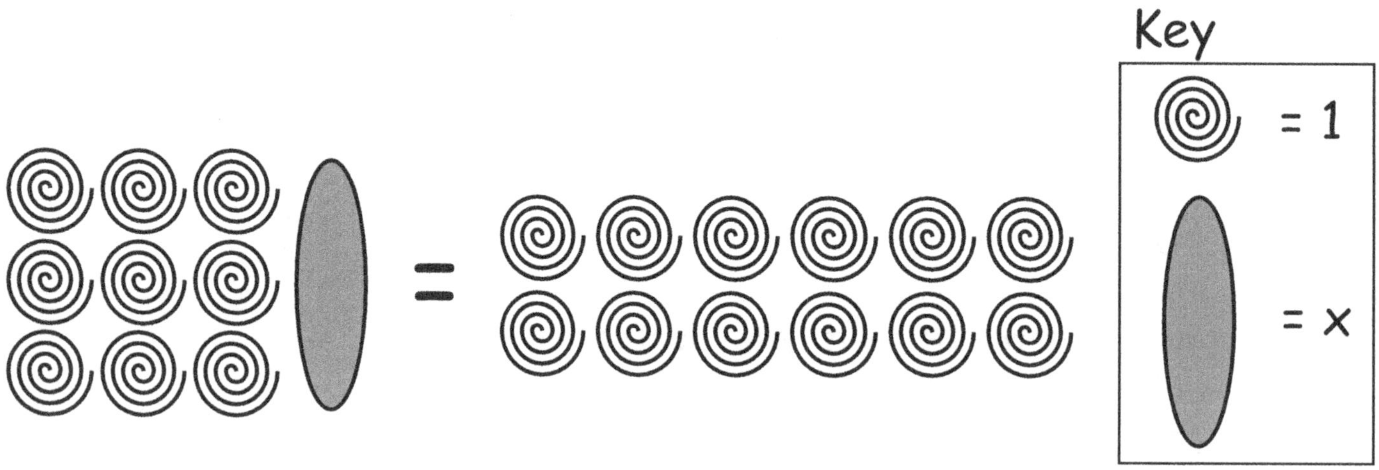

(A) $12x = 9$ (B) $x + 9 = 12$

(C) $12 + x = 9$ (D) $9x = 12$

35. What is the 7th term in the pattern given below ?

40 , 64 , 104 , 168 , 272 ,.....

(A) 831 (B) 440

(C) 712 (D) 721

36. The below pattern is based on a rule. Which of the below options defines the rule ?

243 , 81 , 27 , 9 ....

(A) Divide the previous number by 3
(B) Multiply the previous number by 3
(C) Subtract the previous number by 30
(D) Add 30 to the previous number.

37. The variable in the below expression is ?

$$7y - 42 = 70$$

(A) =
(B) 70
(C) -42
(D) y

38. Find the area of the triangle ABC

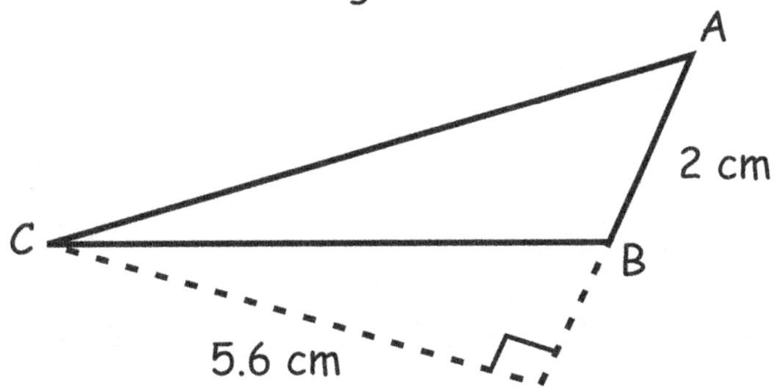

(A) 2.8 cm^2
(B) 5.6 cm^2
(C) 13.1 cm^2
(D) 11.2 cm^2

39. How many ounces are equal to 7 cups ?

(A) 58 (B) 56

(C) 63 (D) 21

40. Find the missing angle in the given quadrilateral ABCD as given below. What is the measure of the ∠x

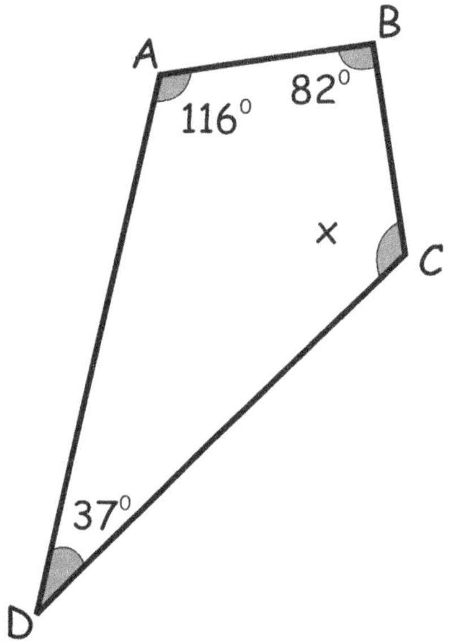

(A) 133° (B) 141°

(C) 131° (D) 125°

41. To find which of the below we need the average of middle values of the data set which has the even number of records ?

(A) Median (B) Range

(C) Mean (D) Mode

42. Find the area of the below circle ? (Approximately)

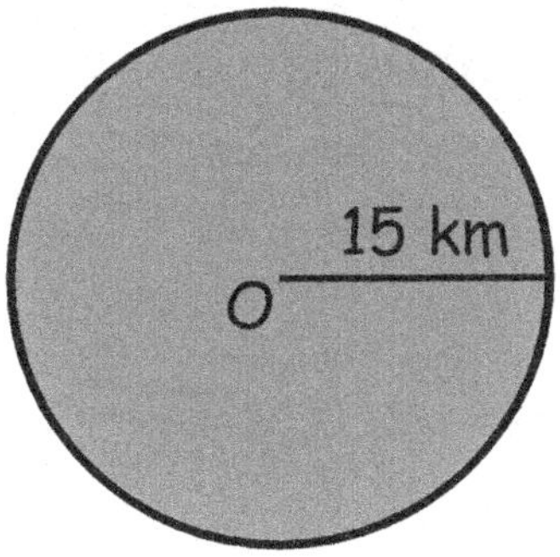

(A) 47 km^2 (B) 11250 km^2

(C) 707 km^2 (D) 188 km^2

43. Which of the below represents all the factors of 66 ?

(A) 1 , 2 , 11 , 22 , 33 , 66
(B) 1 , 2 , 3 , 6 , 11 , 22 , 33 , 66
(C) 1 , 2 , 3 , 6 , 11 , 33 , 42 , 66
(D) 1 , 66

44. What is the least common multiple of

35 , 10

(A) 70 (B) 5

(C) 10 (D) 350

45. Find the median of the below data set

31 , 39 , 39 , 45 , 47 , 49 , 49 , 66 , 66

(A) 57 (B) 53

(C) 60 (D) 47

46. Evaluate $2 \times 4\frac{4}{5} \times \frac{1}{2} = ?$

(A) $4\frac{18}{35}$ (B) $7\frac{3}{10}$

(C) $5\frac{5}{6}$ (D) $4\frac{4}{5}$

47. Find the range of the below data set

64 , 51 , 68 , 57 , 57 , 44 , 51 , 49 , 54 , 56 , 61

(A) 22 (B) 17

(C) 25 (D) 23

48. Find the net of the 3D figure given below

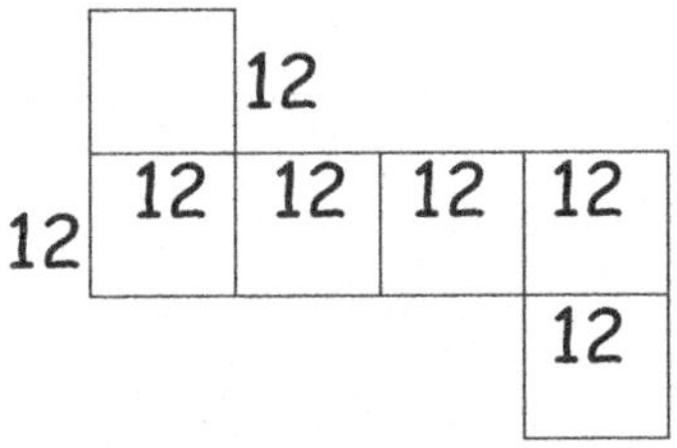

(A) 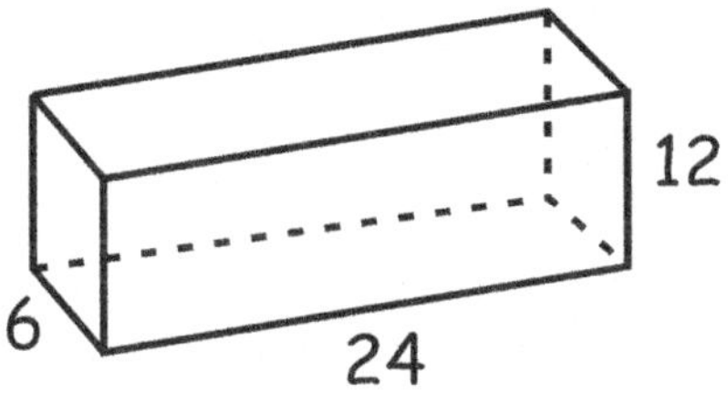

(B) 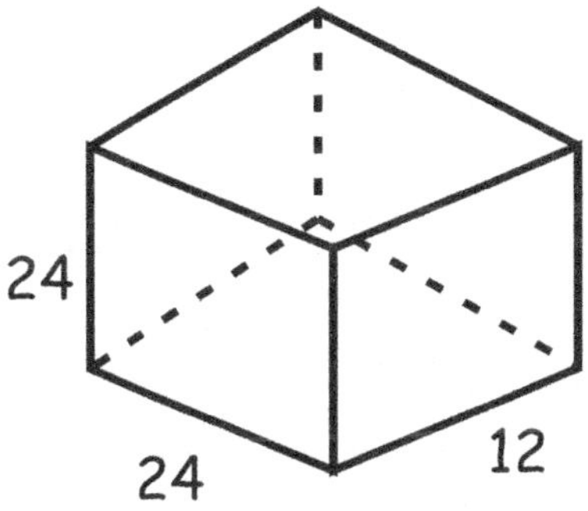

(C) 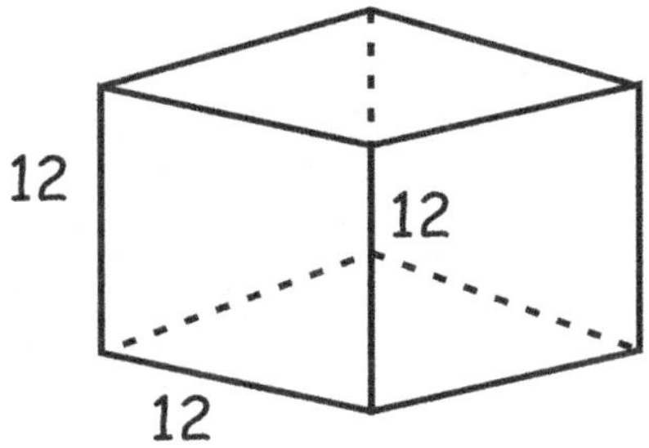

(D) 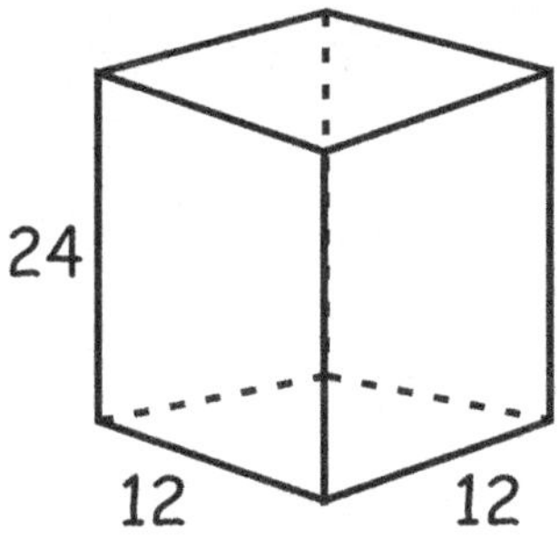

49. Identify the figure below

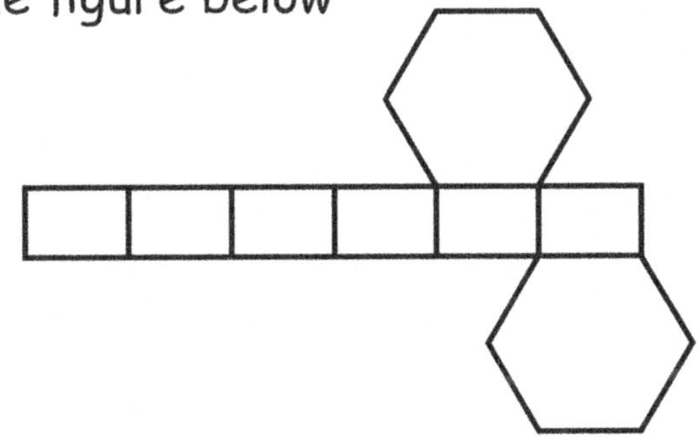

(A) Hexagonal prism

(B) Square prism

(C) Rectangular pyramid

(D) Pentagonal prism

50. There are 4 students running for student government president. A survey was taken asking 100 students which candidate they would vote for in the election. The result are shown in the table below :

Candidate's Name	Number of supporters
Ashley	30
Britney	28
Lisa	14
Walker	28

What is the probability that Lisa will win the election

(A) $\frac{3}{10}$

(B) $\frac{7}{25}$

(C) $\frac{7}{50}$

(D) $\frac{43}{50}$

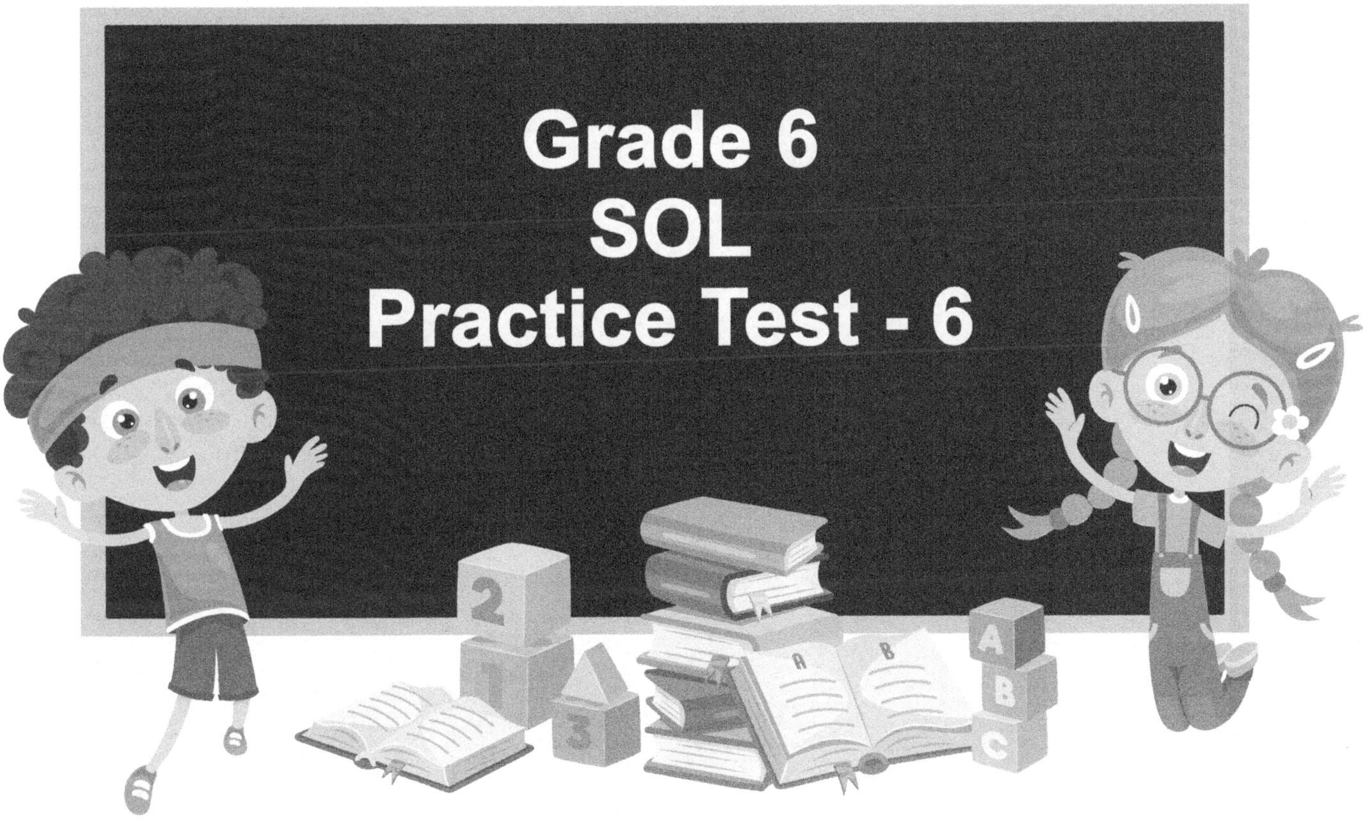
Grade 6
SOL
Practice Test - 6

1. Mat sold type-A town houses for $445,500 and type-B town houses for $865,000 this week. What information is needed to find this week's house sales amount by Mat ?

(A) The number of type-A town houses sold
(B) The number of type-B town houses sold
(C) The number of type-D town houses and type-A town houses sold
(D) The number of type-B town houses andtype-A town houses sold

2. A soccer ball costs $17.34, A basketball costs $19.99, and Football costs $15.90 including sale tax. Andy gave the clerk $100.00. Which is the closest amount Andy has after spending for the three balls?

(A) $46.71 (B) $45.79

(C) $46.77 (D) $44.47

3. Evaluate $97.035 \div 0.5 = ?$

(A) 194.07 (B) 1904.07

(C) 190.47 (D) 1940.07

4. Natt and Neel biked $7\frac{3}{5}$ miles and $5\frac{4}{5}$ miles on the trial this morning. How many more miles Natt biked than Neel on the trial?

(A) $3\frac{1}{5}$ (B) $1\frac{1}{5}$

(C) $3\frac{1}{4}$ (D) $1\frac{4}{5}$

5. Kate buys a back pack for \$101.70 , a calculator for \$66.80 and three sets of markers for 9.99 each. Which is closest to the total amount he got through sales today ?

(A) \$ 199 (B) \$ 179

(C) \$ 190 (D) \$ 178

6. Evaluate $26.2 \div 0.8 = ?$

(A) 28.6 (B) 25.4
(C) 32.75 (D) 22.21

7. Evaluate $\frac{5}{6} \div \frac{3}{14} = ?$

(A) $\frac{1}{11}$ (B) $3\frac{8}{9}$

(C) $1\frac{1}{21}$ (D) $\frac{9}{13}$

8. What is the greatest common factor of

72 , 66

(A) 6 (B) 792

(C) 2 (D) 18

9. A sports store sold a number of items in January 2019 as given below.

ITEM	Number of sports items sold	Cost per item
Bike	105	$ 89
Scooter	90	$ 124.99
Hoover board	75	$ 249.99

Based on the data in the table above, Which could be used to determine the total amount of sales made before tax ?

(A) Divide the cost per item by the number of items, then multiply the quotients

(B) Multiply the cost per item by the number of items, then multiply the product

(C) Multiply the cost per item by the number of items, then add the products

(D) Divide the cost per item by the number of items, then add

10. What is the percentage equivalent of $\frac{9}{25}$?

(A) 360% (B) 36%

(C) $9\frac{1}{4}$ % (D) $\frac{9}{25}$ %

11. Which of the below is the composite number ?

(A) 673 (B) 105

(C) 541 (D) 701

12. Angle measure closest to 160^0

(A) 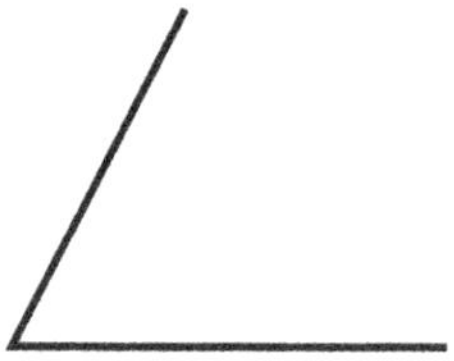

(B) 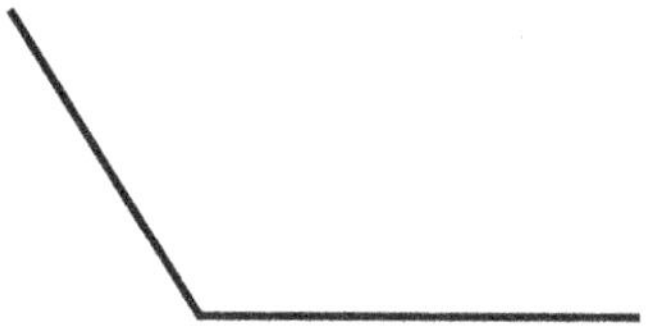

(C) 

(D) 

13. Which of the below statements is true ?

(A) $55 < 81$ (B) $-666 < 666$

(C) $-90 < -170$ (D) $-551 > 330$

14. Composite number is ?

(A) Units digit is always 0.
(B) Not a prime number
(C) Always divisible by 9
(D) Units digit is even number

15. Which decimals are ordered from least to greatest ?

(A) 1.120 , 1.3 , 1.6 , 1.508 , 1.404 , 1.107
(B) 1.6 , 1.508 , 1.404 , 1.107 , 1.120 , 1.3
(C) 1.107 , 1.120 , 1.3 , 1.404 , 1.508 , 1.6
(D) 1.3 , 1.6 , 1.120 , 1.404 , 1.508 , 1.107

16. $1\frac{1}{4}$ of Pamela's garden have vegetables. $\frac{5}{8}$ of the garden has egg plants and the remaining are tomato plants. How much of the vegetable garden area is covered by tomato plants ?

(A) $\frac{5}{8}$ (B) $4\frac{7}{8}$

(C) $1\frac{5}{8}$ (D) $\frac{11}{12}$

17. How many kilometers are equivalent to 5000 centimeters ?

(A) 0.005 kilometers (B) 0.05 kilometer

(C) 0.07 kilometers (D) 0.5 kilometers

18. Which is closest to the area of a legal size paper ?

(A) 14 x 8.5 square inches (B) 14 x 17square meters

(C) 4 x 4square feet (D) 125 feet

19. A circular park has a diameter of 26 miles. Which is closest to the circumference of the park ?

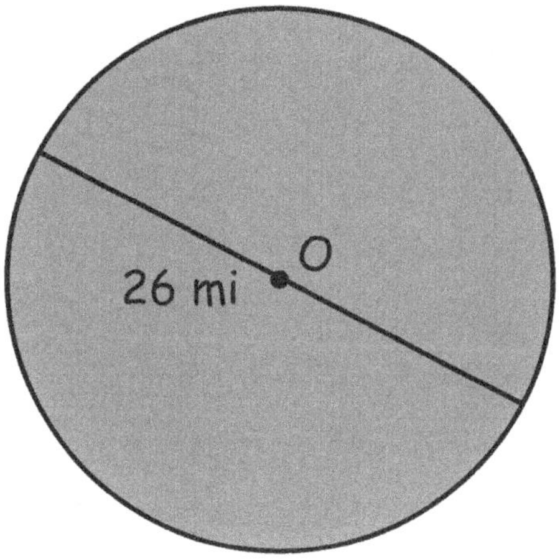

(A) 29∏ mi (B) 8∏ mi

(C) 26∏ mi (D) 27∏ mi

20. What is equivalent to 12 cups ?

(A) 3 pints (B) 4 pints

(C) 12 pints (D) 6 pints

21. What is the probability of getting an odd number when a six-sided dice is rolled ?

(A) $\frac{1}{6}$ (B) $\frac{1}{2}$

(C) $\frac{1}{3}$ (D) $\frac{4}{6}$

22. Find the mean of all the even numbers between 30 and 50. (excluding 30 and 50)

(A) 39 (B) 45

(C) 43 (D) 40

23. 5th grade students of Grace elementary school read a number of books across various weeks over summer vacation. Below is the data collected from various 5th grade sections.

Stem	Leaf
1	2 3 9 9
2	5 5 6 6 6 9
3	8 8 8 8 9
4	0 1 1 2 3
5	0
6	

Key :
1|5 = 15

Which of the following is a *true* conclusion based on the data in the stem-and-leaf plot ?

(A) The highest number of books read in a week are 50.
(B) Five students read 40 books.
(C) Lowest number of books read in a week are five.
(D) Students read same number of books every week.

24. Based on the data given below draw a box and whisker plot. Chose the correct option from the below.

20	23	28	31	34	44	47	64	66

(A)

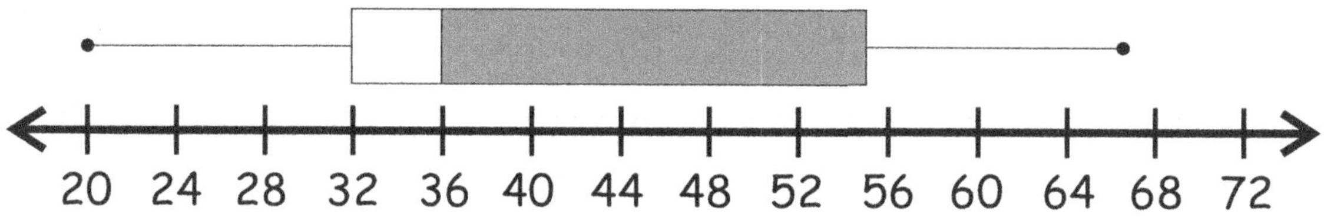

(B)

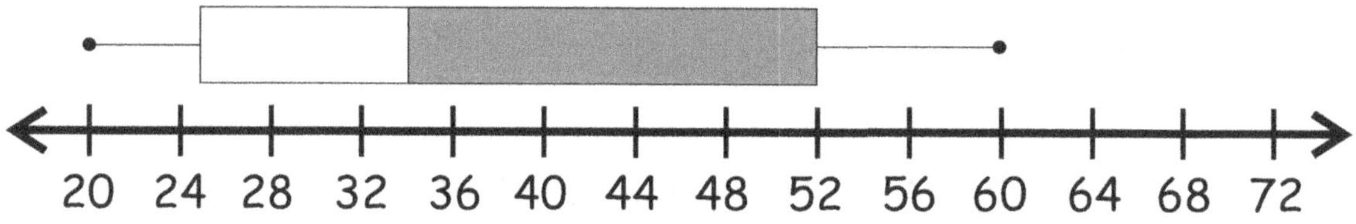

(C)

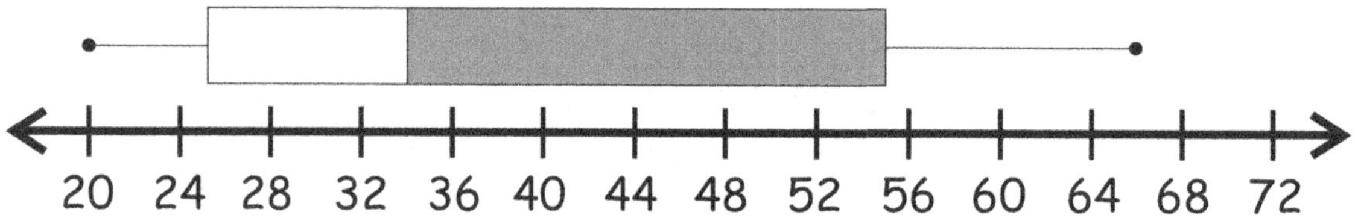

(D)

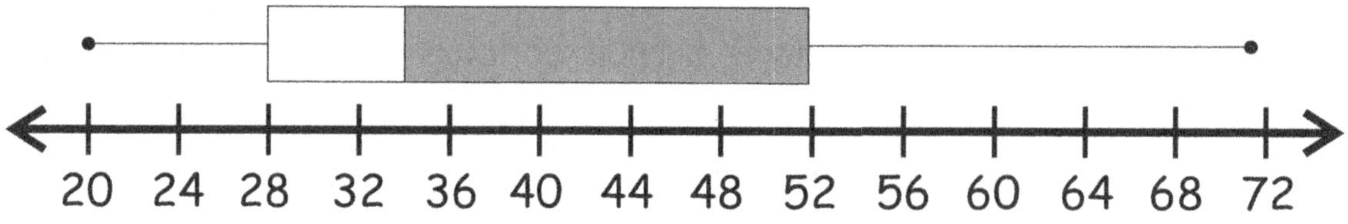

25. Mr. Smith has two types of pants and two types of shoes to chose as per the below data.

Week	Pant color	Shoe color
Monday	Black pant	Black shoes
Wednesday	Blue pant	Brown shoes

Based on the chart, which of the following shows all of Mr.Smith's possible clothing combinations for a given week day ?

(A)

- Sunday
  - Black pant
    - Black shoes
    - Grey shoes
  - Blue pant
    - Black shoes
    - Grey shoes
- Wednesday
  - Black pant
    - Black shoes
    - Brown shoes
  - Blue pant
    - Black shoes
    - Brown shoes

(B)

- Monday
  - Black pant
    - Black shoes
    - Brown shoes
  - Blue pant
    - Black shoes
    - Brown shoes
- Wednesday
  - Black pant
    - Black shoes
    - Brown shoes
  - Blue pant
    - Black shoes
    - Brown shoes

(C)

- Monday
  - Black pant
    - Grey shoes
    - Brown shoes
  - Blue pant
    - Grey shoes
    - Brown shoes
- Wednesday
  - Black pant
    - Black shoes
    - Brown shoes
  - Blue pant
    - Black shoes
    - Brown shoes

(D)

- Wednesday
  - Black pant
    - Black shoes
    - Brown shoes
  - Blue pant
    - Black shoes
    - Brown shoes

26. Find the solution of the below equation ?

$$19 = \frac{x}{11}$$

(A) 30 (B) 8

(C) 1 (D) 209

27. Find the 9th number in the Arithmetic pattern

525 , 500 , 475 , 450 , 425 ,.....

(A) 375 (B) 400

(C) 325 (D) 350

28. Which of the below is not a perfect square?

(A) 576 (B) 441

(C) 783 (D) 484

29. Evaluate the below :

$$18.18^5.18^4$$

(A) 18^{10} (B) 18^2

(C) 18^6 (D) 18^{11}

30. The below number sentences true for what value of "y"

$$y - 26 = -55$$

(A) -2 (B) -81

(C) $\frac{-7}{39}$ (D) -29

31. which word best describes the circled number

$$\textcircled{6}n = 108$$

(A) Variable (B) Co-Efficient

(C) Term (D) Equation

32. Based on the below equation. Find is the value of Z

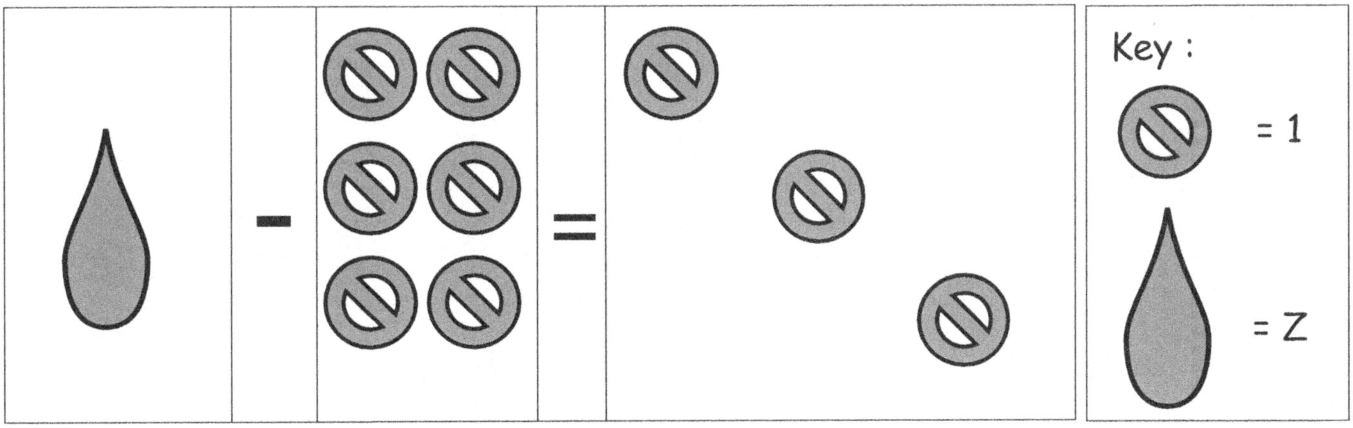

(A) 9 (B) 3

(C) 6 (D) 18

33. What is the 11th figure in the below pattern ?

12121...

(A) 12 (B) 2

(C) 1 (D) 21

34. What number sentence is modeled by the shapes shown ?

(A) $18x = 6$ (B) $18 + x = 6$

(C) $x + 6 = 18$ (D) $6x = 18$

35. What is the 7th term in the pattern given below ?

1165 , 1885 , 3050 , 4935 ,.....

(A) 20905 (B) 12920

(C) 7985 (D) 87

36. The below pattern is based on a rule. Which of the below options defines the rule ?

45 , 405 , 3645 , 32805 , 295245 , ...

(A) Divide the previous number by 9
(B) Multiply the previous number by 9
(C) Subtract the previous number by 90
(D) Add 99 to the previous number.

37. The variable in the below expression is ?

$$\frac{8}{3}z + 66 = 9$$

(A) z

(B) $\frac{8}{3}$

(C) +

(D) 66

38. Find the area of the quadrilateral ABCD given below

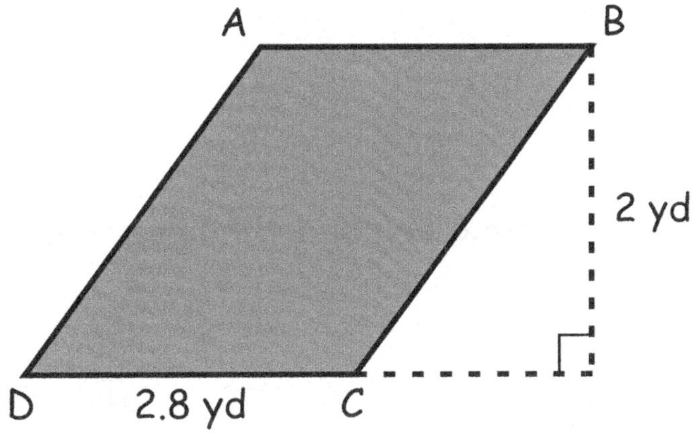

(A) 5.6 yd^2

(B) 11.3 yd^2

(C) 11.2 yd^2

(D) 2.8 yd^2

39. How many pints equal to 6 quarts ?

(A) 18 (B) 16

(C) 12 (D) 24

40. Find the missing angle of the quadrilateral ABCD given below

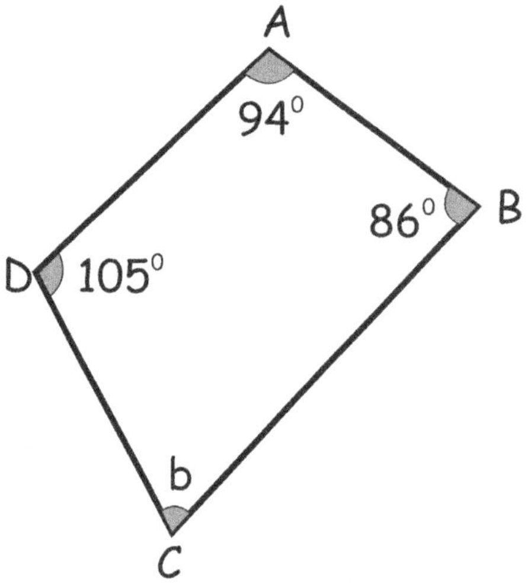

(A) 86^0 (B) 75^0

(C) 80^0 (D) 85^0

41. To find which of the below we need the lower and upper quartile values of the data set ?

(A) Median (B) Range

(C) Average (D) Box and whisker plot

42. Find the area of the circle given below ? (Approximately)

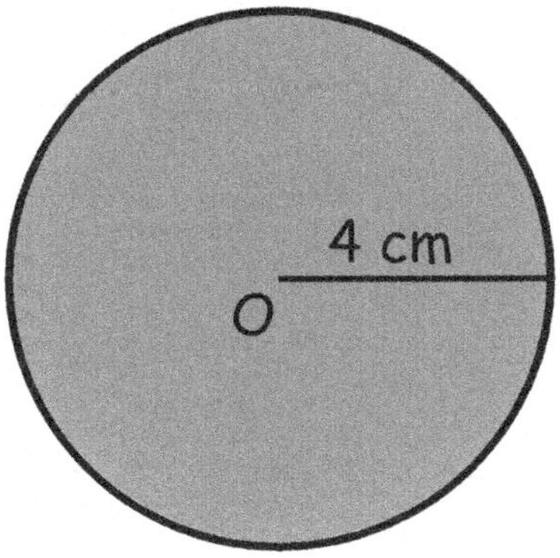

(A) 200 cm^2 (B) 50 cm^2

(C) 1134 cm^2 (D) 210 cm^2

43. Which of the below represents all the factors of 75 ?

(A) 1 , 3 , 5 , 15 , 25 , 75
(B) 1 , 3 , 25 , 75
(C) 1 , 3 , 5 , 15 , 19 , 75
(D) 1 , 3 , 5 , 10 , 15 , 25 , 75

44. What is the least common multiple of

25 , 15

(A) 16 (B) 5

(C) 75 (D) 375

45. Find the median of the data given below

1 , 3 , 5 , 7 , 7 , 7 , 16 , 27 , 30

(A) 5 (B) 7

(C) 4 (D) 6

46. Evaluate $4\frac{1}{2} \times \frac{2}{3} \times \frac{1}{5} = ?$

(A) $\frac{3}{5}$ (B) $4\frac{2}{5}$

(C) $\frac{1}{8}$ (D) $6\frac{1}{2}$

47. Find the range of the below data set

1.5 , 10 , 50 , 3 , 35 , 15 , 20 , 20 , 40

(A) 45 (B) 64

(C) 74 (D) 48.5

48. Identify the figure given below

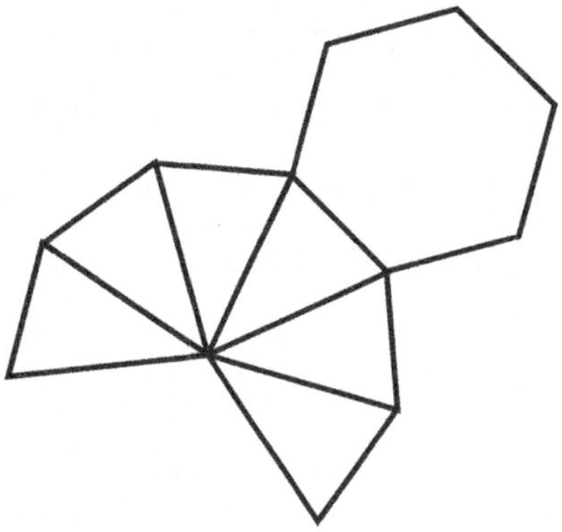

(A) Square prism
(B) Hexagonal Pyramid
(C) Square pyramid
(D) Cylinder

49. Sales of "Brand A" car sales across various years is shown as below

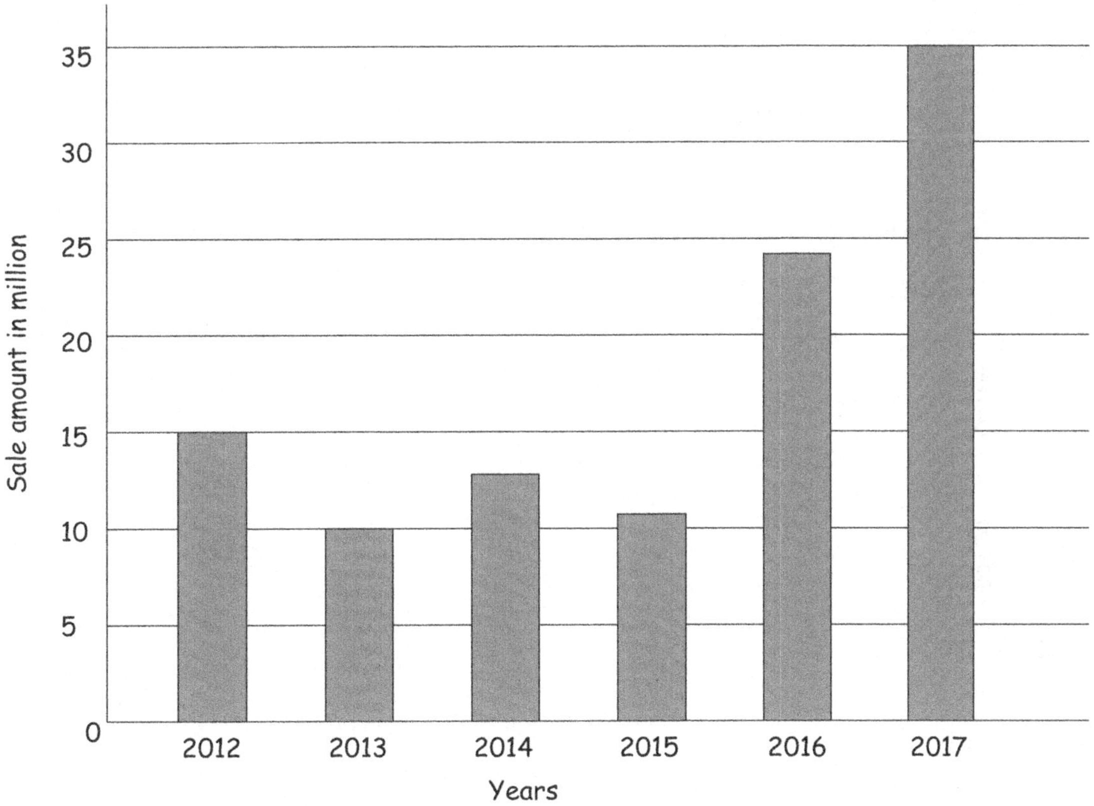

What is the ratio of sales in 2017 to 2012 ?

(A) 5 : 3
(B) 7 : 5
(C) 7 : 3
(D) 3 : 7

50. There are 3 students running for student government president. A survey was taken asking 40 students which candidate they would vote for in the election. The result are shown in the table below : (Few may refuse to vote)

Candidate's Name	Number of students votes
Reese	15
Matthew	13
Chris	12

Based on this table what is the probability that the students will vote for Reese ?

(A) $\frac{1}{3}$

(B) $\frac{3}{5}$

(C) $\frac{3}{8}$

(D) $\frac{5}{8}$

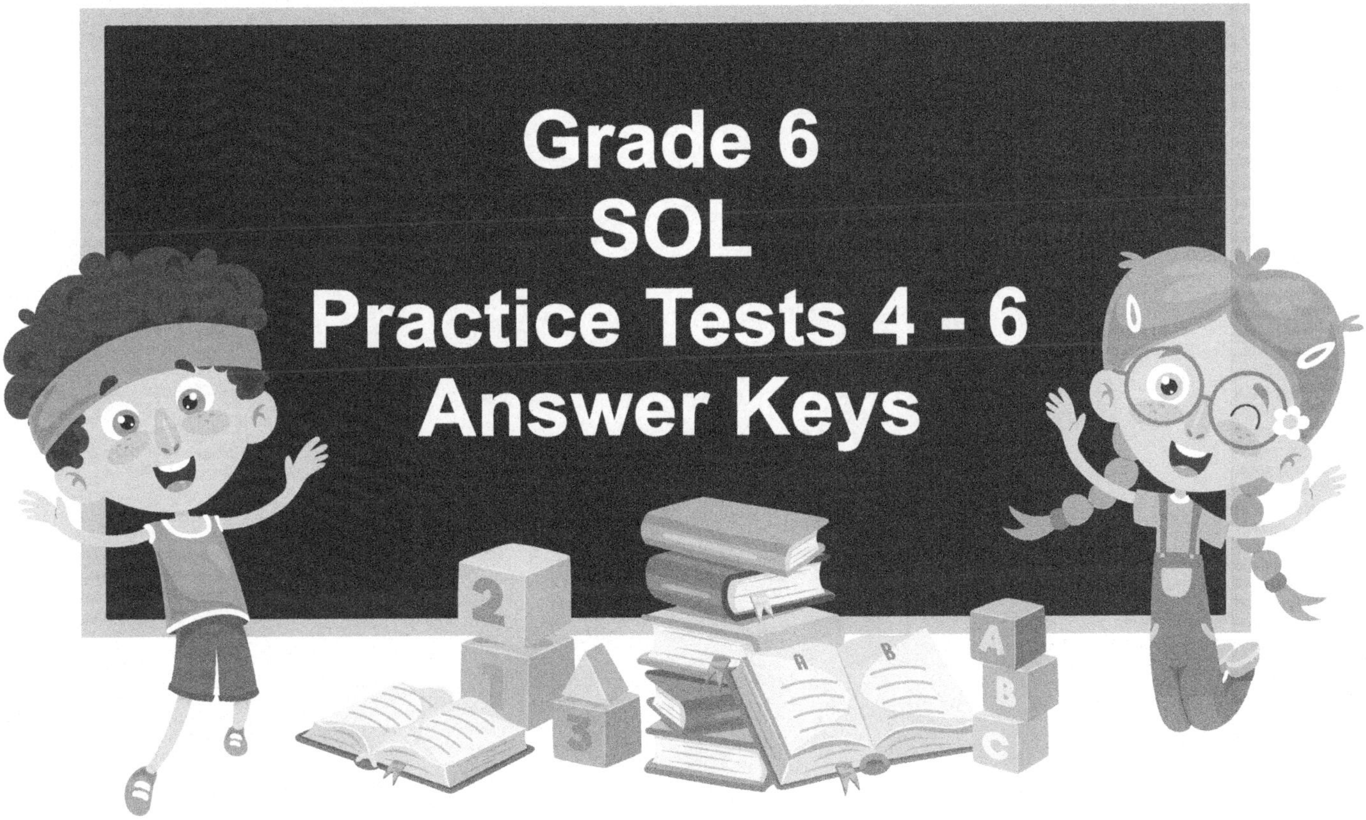
Grade 6
SOL
Practice Tests 4 - 6
Answer Keys

Answer Key Test - 4

1. A
2. C
3. A
4. A
5. A
6. D
7. A
8. B
9. D
10. B
11. B
12. A
13. C
14. B
15. A

16. D

17. B

18. C

19. D

20. B

21. D

22. C

23. A

24. A

25. C

26. D

27. B

28. D

29. B

30. C

31. C

32. D

33. B

34. C

35. A

36. D

37. A

38. D

39. B

40. C

41. A

42. A

43. C

44. D

45. A

46. B

47. D

48. C

49. D

50. B

Answer Key Test - 5

1. D
2. A
3. C
4. A
5. C
6. A
7. A
8. C
9. B
10. D
11. B
12. A
13. C
14. D
15. A

16. C

17. B

18. A

19. D

20. C

21. B

22. C

23. D

24. A

25. C

26. A

27. C

28. D

29. A

30. A

31. C

32. C

33. A

34. D

35. C

36. A

37. D

38. B

39. B

40. D

41. A

42. C

43. B

44. A

45. D

46. D

47. A

48. C

49. A

50. C

Answer Key Test - 6

1. D
2. C
3. A
4. D
5. A
6. C
7. B
8. A
9. C
10. B
11. B
12. D
13. B
14. B
15. C

16. A

17. B

18. A

19. C

20. D

21. B

22. D

23. A

24. C

25. B

26. D

27. C

28. C

29. A

30. D

31. B

32. A

33. C

34. D

35. A

36. B

37. A

38. A

39. C

40. B

41. D

42. B

43. A

44. C

45. B

46. A

47. D

48. B

49. C

50. C

GRADE 6
SOL
Score Calculation

Score calculation

If you get this many times correct :	Then your converted scale scor is :
0	000
1	217
2	249
3	268
4	282
5	293
6	303
7	311
8	319
9	325
10	331
11	337
12	343
13	348
14	353
15	357
16	362
17	366
18	371
19	375

20	379
21	383
22	387
23	391
24	395
25	399
26	403
27	407
28	411
29	415
30	419
31	423
32	427
33	432
34	436
35	441
36	445
37	450
38	456
39	461

40	467
41	473
42	480
43	487
44	496
45	505
46	516
47	531
48	550
49	582
50	600